Pocket Handbook

The essential guide to property and construction, as used by professionals since 1983

First published in 1983, the Watts Pocket Handbook is an annual publication.

The editors wish to express their indebtedness to previous editors of the handbook, who laid the foundations on which this 19th edition has been built.

The editors would like to acknowledge with thanks the contributions of individuals in the property market and construction industry who send suggestions and comments. Such contributions are always welcome and should be sent to Samantha Rumens, Communications Executive, Watts and Partners, 1 Great Tower Street, London, EC3R 5AA.

No guarantee is given by Watts and Partners or any of its employees as to this Pocket Handbook's accuracy. More detailed advice must be obtained before relying on statements made. Library and research facilities are available to assist in developing particular lines of enquiry.

Published by Watts and Partners
Typeset and printed by Cantate Bishops
Cover design work by Art and Industry

Contents

Introduction   1

Property acquisition and disposal   3
   Commercial and industrial property   3
      Commercial/industrial surveys   4
      Institutional standards   6
   Housing and residential property   7
      Residential surveys   8
   Due Diligence   11
      Due diligence and the building survey   12
      Vendor's surveys   14
      Development monitoring   14

Development and procurement   17
   Contracts and procurement   17
      Procurement methods   18
      Procurement and standard form contracts   18
      The contract administrator's role   20
   Contract management   21
      Employer's agent   22
      The quantity surveyor's role   23
      Partnering   23
   Development   25
      Construction management   26
      Project management   27
      Quality management and professional construction services   27

Legislation and regulation   29
   Building legislation and control   29
      Acts of parliament and regulations   30
      The Fire Precautions Act 1971   33
      Fire Precautions (Workplace) Regulations 1997 and
      (Amendment) Regulations 1999   34
      The Disability Discrimination Act 1995   36
   Building and construction regulations   41
      The Building (Amendment) Regulations 2001   42
      Airtightness   43
      The Provisions of Part II Housing Grants, Construction and
      Regeneration Act 1996   44

Contents

	Health and safety at work	48
	Construction (Health, Safety and Welfare) Regulations 1996	49
	The Construction (Design and Management) Regulations 1994	51
Town and country planning in England		**55**
	Development applications and fees	56
	Appeals and called-in applications	56
	Development plans and monitoring	56
	General Development Order and Use Classes Order	57
	Planning policy guidance notes and circulars	58
	Listed buildings and conservation areas	59
	Environmental impact assessments	61
Legal and lease		**63**
	Dilapidations	64
	Latent Damage Act 1986	70
	Discovery of building defects - statutory time limits	71
	Expert witness	71
	Dispute resolution	74
	Adjudication under the Scheme for Construction Contracts - how to get started	76
Neighbourly matters		**79**
	Rights to light	80
	Daylight and sunlight	85
	Party wall procedure	86
	Access agreements	89
	Construction noise and vibration	90
Architectural, engineering and services design		**91**
Architectural and design criteria		**91**
	Basic design data	92
	Building types	93
	Specifications	96
	Different specifications for different contracts	96
	Specification writing	96
	Site archaeology	97
	Coordinating project information	101
Structural and civil engineering design		**103**
	The design process	104
	Soils and foundation design	104

Contents

Soil survey	105
Design loadings for buildings	108
Building services design	**115**
Air conditioning systems	116
Plant and equipment	119
Lift terminology	120
Lighting design	121
Data installations	122
Site analysis	**123**
Measured surveys	124
Computer aided design	**127**
Computer aided design	128
Tables and statistics	**131**
Conversion formulae	132
Materials and defects	**133**
Materials	**133**
Deleterious materials	134
Asbestos	136
High Alumina Cement concrete	141
Chlorides	143
Corrosion of metals	145
Building defects	**147**
Common defects in commercial properties	148
Common defects in residential properties	149
Problem areas with 1960s buildings	150
Defects in concrete	159
Fungi and timber infestation in the UK	160
Rising damp	165
Rising groundwater	167
Testing	**169**
Chemical and physical testing requirements	170
Non-destructive testing	171
Cladding	**173**
Curtain walling systems	174
Mechanisms of water entry	178
Glazing - windows and doors satisfying the Building Regulations	179
Spontaneous glass fracturing	181

Contents

	Conservation and the environment	**183**
	Contaminated land	184
	Energy conservation	186
	Environment and specification	188
	Radon	189
	BREEAM	191
	Maintenance management	**193**
	Primary objectives	194
	A systematic approach	194
	Condition surveys	197
	'Best value' in local authorities	199
	Sources of information in maintenance management	201
	Cost management and taxation	**203**
	VAT in the construction industry - zero rating	204
	Capital allowances for taxation purposes	207
	Tender prices and building cost indices	210
	Cost management	211
	Life cycle costings	212
	Reinstatement valuations for insurance purposes	216
	Useful information	**219**
	Useful information sources	220
	Useful website addresses	221
	Contributors to the Watts Pocket Handbook	225
	Watts and Partners' publications	228

Index 229

Watts and Partners offices 238

Introduction

Are you looking for an up-to-date, technical reference on property and construction? If so, look no further than the Watts Pocket Handbook - the essential guide to property and construction, as used by professionals since 1983.

In this, the 19th edition, you will find the same level of expertise that has made the Handbook so successful in previous years. The content, which covers both managerial and technical issues, has been updated to reflect change and progression in legislation and work practices in the last year. The easy-to-use format offers you rapid access to practical information.

There are several new sections this year addressing all stages of the property cycle: from asbestos to development monitoring; and from procurement to vendor's surveys. We believe all information and legal references to be correct at the time of going to press. However, it is advisable, in specific instances, to obtain appropriate professional advice.

We welcome your contributions and comments, as we strive to fulfil our mission to publish the best information source for surveyors and other property and construction industry professionals. The 2003 issue continues to build on the high standards of past editions, making the Handbook a recognised industry standard - something that we, as a firm, are very proud of.

Peter Primett
Chairman
Watts and Partners

Commercial and industrial property

 Commercial/industrial surveys

 Institutional standards

Property acquisition and disposal

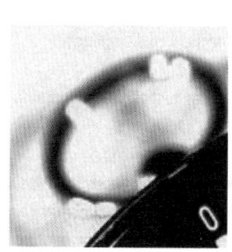

3

Property acquisition and disposal

Commercial/industrial surveys

Check lists for establishing brief and extent and method of survey:

Identify the reason for and scope of survey

- acquisition;
- general or specific;
- adaptation, extension, change of use;
- maintenance programme, expenditure plan;
- value as an investment;
- insurance assessment; and
- investment or occupation.

Ascertain the degree of detail required

- delivery, access;
- floor loadings, gantry loadings;
- security;
- fire protection;
- means of escape;
- environmental control and other matters;
- special processes, dangerous or toxic materials; and
- degree of opening up of concealed areas.

Identify energy conservation requirements

- use of specialists for inspecting, testing, etc.

Agree extent of tests/analysis of materials

- asbestos (type, content, dust counts);
- High Alumina Cement;
- calcium chloride;
- carbonation;
- covermeter or other non-destructive tests;
- methane gas;
- nature of core materials in composite panels;
- quality/constituents;
- extent and method of sampling;
- making good; and
- extent of services tests including drainage.

Ascertain access for inspection

- ladder, cradle, hydraulic hoist, abseil, specialist help required;
- public liability insurance and insurance of hire equipment; and
- special access requirements for confined spaces.

Property acquisition and disposal

Ascertain tenure and request relevant documents
- freehold/leasehold;
- copies of leases, deeds and any plans attached; and
- advice on effects of repairing covenants, schedules of condition, rebuilding, reinstatement clauses.

Contaminated land
- possible contaminative uses (but check that professional indemnity (PI) insurance covers this risk).

Ascertain information available
- history;
- base building specification;
- original architects, engineers, developers, builders;
- floor plans, other drawings, details;
- maintenance records;
- maintenance personnel;
- collateral warranties;
- building contracts/status of works;
- health and safety file; and
- fire certificate.

Consider impact of regulations
- Disability Discrimination Act;
- Workplace Regulations:
    - fire
    - glazing
    - sanitary provisions
    - lighting
    - protection from falling;
- Construction (Design and Management) (CDM), health and safety; and
- means of escape.

Establish if costings are required
- approximate for budget purposes; and
- state limitations on costing information.

Establish and confirm restrictions
- access arrangements;
- restricted, high security areas;
- means of identification, security arrangements;
- potentially hazardous areas;

Property acquisition and disposal

- normal working hours/weekend working; and
- need to undertake risk assessment.

When surveying property – especially when empty – practitioners should conform to the procedures outlined in 'Practical Procedures' under the 'Health and safety at work' section of this handbook.

Institutional standards

The term 'institutional standards' is often referred to in connection with development work or the acquisition of property investments by financial institutions.

In practice it is difficult to define the term precisely as standards will vary between individual investors, location, market factors and the like.

The philosophy of various investors will vary widely and standards that may be unacceptable to one organisation may be perfectly acceptable to another.

The following list represents some of the more important standards, but it should be remembered that failure to meet any one of the items is not in itself sufficient to condemn a property. The institution will need to form a value judgement and it is the surveyor's duty to provide clear, factual information to assist in this. For example, a building in a poor state of repair may be an insignificant factor if the real value of the investment is in the site. As a fundamental rule however, investors require liquidity of assets – the more 'warts', the narrower the market.

The standards are listed below:-

- Freedom from deleterious materials or systems that may be prejudicial to health (for example, wet cooling towers, Legionnaires' disease).
- Compliance with statutory requirements, health and safety requirements, building regulations, fire precautions and means of escape and planning.
- Fitness for purpose. Constructed to a standard and quality generally acceptable for style of premises and locality. Accommodation to allow for tenants changing special requirements or possible long-term refurbishment.
- Compliance with design codes of practice, for example, floor loadings (in some cases institutions will require enhanced standards, for example, 5 kN/m^2 for office accommodation, although there is a trend towards less onerous standards here).
- Freedom from major structural faults or severe maintenance problems.
- Freedom from particular risks in ground conditions: for example, waste chemicals, methane gas, mining subsidence, etc.
- Good title.
- Good covenant.

Investors are beginning to show an increased awareness of environmental issues – CFCs, tropical hardwoods, refrigerants, solvents, etc and due consideration to these issues should be given.

Housing and residential property

 Residential surveys

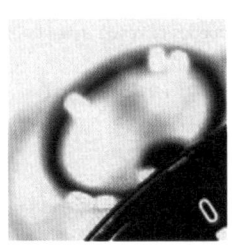

Property acquisition and disposal

Property acquisition and disposal

Residential surveys

Suggested pre-survey checklist for surveyors:

Confirm instructions
- nature of the instructions;
- date and time of the survey;
- access arrangements, particularly occupied premises;
- statement of surveyor's intentions; and
- limitations.

Bear in mind
- Unfair Contract Terms Act 1977;
- liability in negligence and contract; and
- "requirements of reasonableness".

Equipment to take – consider the following
- powerful, robust torch;
- claw hammer and bolster;
- ladder (minimum 3 metres long);
- pocket probe;
- binoculars or telescope (not more than x8);
- hand mirror (minimum 100mm x 100mm);
- moisture meter;
- screwdrivers – assorted;
- measuring rods or tapes, notebook and writing equipment;
- plumbline;
- spirit level;
- first aid kit; and
- protective clothing, hard hat and suitable footwear.

Site notes to be recorded
- notes of defects;
- record of weather, persons present, etc; and
- answers to queries of vendor/neighbours, etc.

Suggested outline of report
Introduction
- brief;
- limitations;
- general description of property and situation; and
- present accommodation.

Property acquisition and disposal

Structural condition and state of repair

External
- roofs;
- other defects at roof level;
- eaves;
- flashings;
- external walls;
- likelihood of cavity wall tie failure;
- airbricks;
- damp proof course;
- foundations and settlement/subsidence;
- rainwater goods;
- soil/waste stacks, gullies; and
- other external comments.

Internal
- roof spaces;
- partitions;
- plasterwork;
- windows;
- doors;
- joinery;
- ceilings; and
- other internal defects.

Services
- plumbing/wiring;
- heating;
- electricity/gas;
- sanitary fittings; and
- drainage.

Outside
- boundaries, pavings, fences, gates and outbuildings;
- noise, contamination and other environmental factors;
- adjoining properties;
- garden/trees – risk of ground movements/mining subsidence;
- garage/car parking; and
- probability of flooding.

General
- compliance with statutory regulations;
- decorations – external and internal;
- planning situation;
- any responsibilities under a lease;

Property acquisition and disposal

- ❖ limitations of report;
- ❖ any appropriate approximate costings;
- ❖ presence/condition of toxic materials, for example, asbestos;
- ❖ general condition and conclusion;
- ❖ risk and likelihood of fungal decay, insect infestation or conditions that could give rise to these attacks;
- ❖ risk of contaminated land;
- ❖ risk of Radon emissions, power lines, etc; and
- ❖ any other relevant information.

Golden rules

When drafting the report, endeavour to avoid technical jargon but take care to communicate precisely.

Avoid assumptions and ill-thought out statements.

When describing elements answer the following questions:-

1. What is it?
2. What is wrong with it?
3. What will need to be done to put it right?
4. What are the consequences of not putting it right?

Health and safety

When surveying a property – especially one that is empty – practitioners ought to conform to the procedures outlined in 'Practical Procedures' under the 'Health and safety at work' section of this handbook.

See also 'Surveying Safely' published by The Royal Institution of Chartered Surveyors.

Due diligence

 Due diligence and the building survey

 Vendor's surveys

 Development monitoring

Property acquisition and disposal

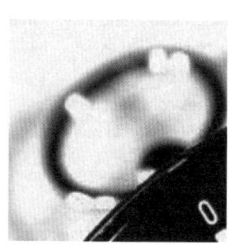

Property acquisition and disposal

Due diligence and the building survey

The building survey is just one part of the process of property acquisition; its significance should not be underestimated. While it may be true to say that the decision to purchase or occupy is often governed more by commercial pressures than by faults in the building, the due diligence process is designed to alert the purchaser to issues that will affect the building as an investment and or as an asset – to manage the risks that are inherent in property acquisition. The 'conventional' survey may therefore be expanded to encompass a wide range of issues that might, in the normal course of events, be disregarded.

There is no hard and fast rule as to what should be included in the process and what should not; the particular nature of a deal or building will dictate particular investigations. However, the following issues may be relevant:

Occupational considerations

- Constructional issues relating to fit out and occupation
- Suitability of space for particular use
- Ease of sub-division/sub letting
- Contributions from landlord
- Efficiency of space usage
- Quality of base build information and accommodation
- Oversized/undersized
- Access
- Parking restrictions
- Security
- Service charge levels - past history
- Workplace Regulations (fire, lighting, glazing, sanitary provision, protection against falling)
- Disability Discrimination Act
- Occupancy cost review

Repairs and defects

- Patent defects
- Potential latent defects
- Compliance with statute
- Schedules of Condition (and effects thereof)
- Onerous maintenance issues
- Deleterious materials
- Quality of finishes/construction
- Unusual construction techniques

Environmental considerations

- Risk of contamination
- Previous site use

Property acquisition and disposal

- ❖ Remediation
- ❖ Power lines
- ❖ Japanese Knotweed
- ❖ Flooding
- ❖ Radon
- ❖ Coast erosion
- ❖ Mining
- ❖ Security

Legal issues

- ❖ Boundaries
- ❖ Services crossing boundaries
- ❖ Party wall issues
- ❖ Rights to light
- ❖ Restrictive covenants
- ❖ Planning
- ❖ Means of escape over adjoining land
- ❖ Adjoining uses
- ❖ Warranty package
- ❖ Building contract issues
- ❖ Defects liability
- ❖ Title
- ❖ Repairing obligations
- ❖ Lease break provisions
- ❖ Reinstatement obligations
- ❖ Dilapidations liability
- ❖ Fire certification

Due diligence team members

- ❖ Building surveyor
- ❖ Services engineer
- ❖ Environmental engineer
- ❖ Agent
- ❖ Managing agent
- ❖ Solicitor
- ❖ Planner
- ❖ Space planner
- ❖ Structural engineer
- ❖ Cladding specialist
- ❖ Project manager
- ❖ IT consultant
- ❖ Cost consultant

Property acquisition and disposal

Vendor's surveys

The traditional approach to selling investment grade property is to release details to the market, consider offers, reach heads of terms which are often subject to survey and legal enquiries. This due diligence process may give the prospective purchaser cause to renegotiate terms with a risk to the vendor in terms of cost and time. If matters proceed smoothly, exchange of contracts and completion can then take place.

The process is time consuming and fraught with risk. To streamline the procedure and manage risks, vendors are increasingly procuring full survey reports prior to sale – the government's 'Sellers Pack' for the residential market is a similar concept. Vendor's surveys can be defined as building and other surveys commissioned by a vendor but primarily for the benefit of a purchaser.

Normally, the vendor's sale pack will include the normal building survey, a phase 1 environmental report or land quality statement, test reports on deleterious materials (where relevant) and probably a report on the building's services installations.

Whereas a conventional report may often make recommendations for further investigation, a vendor's survey must not invite further questions. Thus, it is very important to either make a judgement and express an opinion based upon the evidence, or commission additional tests and inspections where it is relevant to do so. Similarly, questions that would normally be referred to the legal team should be addressed in advance of the production of the final report. In other words, every effort must be made to 'close' particular issues or observations.

Traditionally, building and other surveys include a number of limitations. The third party clause (where the report can only be relied upon by the client) clearly needs variation with vendor's surveys. There is always some scope for discussion, but the usual basis is that the client (or vendor) can rely upon vendor's survey reports as well as the first purchaser. A duty is often extended to the first purchaser's bankers. Furthermore, there should be no change of use, as this may have impacts on the building that we have not foreseen. The purchaser must also accept that the building's condition may have changed since the date of our report. Assignment of the report to the first purchaser is traditionally done by exchange of letters but can also be executed as a deed.

Advantages of vendor's surveys include making information on the condition of the property available to the vendor, which is essential for good asset management. The surveys also take technical due diligence off the critical path, because they are prepared before the property is marketed. Heads of terms can be entered in to without being subject to survey. This approach may also give less scope to prospective purchasers who are not sincere in proceeding on the basis of their original offer.

Development monitoring

Development monitoring is distinct from both project management and construction monitoring and can be defined as:

> The means of identifying, monitoring and controlling the risks associated with acquiring an interest in a development or entering into an agreement to purchase or let a completed development, which is not under the client's direct control.

Development monitoring is often carried out on behalf of a funding institution that will acquire the scheme as an investment on completion, it might also provide funding during the construction phase. However, development monitoring is also undertaken by financiers when the loan expires at the end of a development period and they therefore have no

Property acquisition and disposal

interest in the completed scheme. It may also be carried out on behalf of occupiers who enter into a commitment to lease a property before completion and even on behalf of insurers where there is to be a decennial policy.

Acquiring an interest in a development at any time before completion at either schematic design stage or once construction has commenced, entails risk. The agreement between parties will seek to apportion this. These risks fall under two headings: a change in value of the scheme between commitment and completion; and the risks associated with construction.

- ❖ Values will change when rents fall (or rise) and when yields vary.
- ❖ Construction risks are myriad and range from an inappropriate procurement route or professional team structure to poor management of the project leading to a risk in construction costs.

Interests can be mitigated, but inevitably some will remain and the agreement between parties should identify and apportion the risk.

Monitoring service

There is no standard development monitoring service, as the nature of the appointment will vary to suit each building and the interest being acquired in it. In general terms, the surveyor or engineer will comment on the performance, programming and cost criteria from the initial concept through to establishing a maintenance regime for the completed scheme.

Development monitoring is usually carried out in four stages:

1. Appraisal and risk assessment.
2. Construction and finance monitoring.
3. Involvement at completion.
4. Involvement post-completion (during and at the end of the defects liability period).

Appraisal and risk assessment

The initial report comments on a wide range of topics including programme, cost, an audit of the design considering durability and compatibility of materials, practical considerations of build or buildability and compliance with both statutory requirements and expectations of occupiers and investors. It deals with the suitability of the project structure, from procurement route to the selection of professionals and contractors involved.

Understanding where the 'focus' of this report should be depends upon the relationship between the parties. The document setting out this relationship is prepared by solicitors but the monitoring surveyor can provide essential input to assist in translating the legal jargon into what is practical on a particular development.

Construction and finance monitoring

The continuing monitoring involves periodic site inspections and attendance at site meetings. Depending on the client's requirements, formal reporting is usually carried out on a monthly or fortnightly basis. These reports comment on a variety of issues including the following:

- ❖ construction costs, cashflow and expenditure;
- ❖ progress of works against programme;
- ❖ quality of workmanship on site;

Property acquisition and disposal

- ❖ development of design;
- ❖ status of any statutory approvals; and
- ❖ status of appointment documents and warranties.

Involvement ante-completion

Within the agreement between parties, there would normally be specific criteria for this stage in the development. The monitoring surveyor represents the client's interests and this often becomes contentious especially where it acts as a trigger for rent commencement or payment of a developer's profit. The brief usually continues to ensure that health and safety files and operation and maintenance manuals are obtained. The monitoring surveyor also needs to ensure that correct procedures are in place to remedy any defects or 'snagging'.

Involvement post-completion

This stage largely focuses on concluding matters, ensuring that all relevant information is in place and that any defects arising during the defects liability period are properly made good.

Advantages of project monitoring

Despite the risks, project monitoring provides a number of significant advantages for both the developer and the party acquiring an interest in the project. As far as the developer is concerned, pre-letting or obtaining finance reduces uncertainty, although this will be reflected by a reduction in profit.

As far as the party acquiring the interest is concerned, advantages include the ability to influence the scheme (which is not possible when acquiring a finished building) and because there is an element of risk, the building is often available on better terms. In a more buoyant market, this may be the only way to acquire the right kind of property in the right location.

Clients value the support and advice which development monitoring brings by identifying the risks and benefits that are inherent in any scheme.

Contracts and procurement

 Procurement methods

 Procurement and standard form contracts

 The contract administrator's role

Development and procurement

Procurement methods

Essentially the function of construction contracts is to assign appropriate levels of risk to those parties best able to deal with them.

Risk is an inherent element of any construction project and therefore risk management is an essential part of contract strategy.

There are currently three main procurement options used within the industry which reflect various ways by which risk is balanced between the parties.

Traditional procurement

A traditional procurement route may be defined as one where the design is largely complete before either the main contractor, sub-contractor or specialist contractors become involved. This may be appropriate for some projects if the client's objectives have been clearly and comprehensively determined. However, it does carry with it the disadvantages of increased economic uncertainty and limited opportunities to refine design and improve cost efficiency. It is also possibly more likely to lead to disputes. Increasingly, when traditional contracts are used, the contractor is selected on a two stage basis to gain some of the advantages that are available from early contractor involvement.

Design and build procurement

One alternative to traditional contracting is design and build procurement, where part or all of the design development and financial outcome responsibility risk is delegated to the constructor. There are many considerations that need to be carefully evaluated in choosing this procurement route, and opinion is divided as to whether it should be used on all types of construction projects. Most commonly this route is used in the procurement of 'spec' buildings, such as industrial units or office buildings when the control of design criteria and building techniques is not a dominant factor for the client.

Management procurement

The key feature of any management contract is that a manager is appointed with the responsibility to manage a project, not just provide advice or consultancy services. Two main derivatives of this form are management contracts and construction management.

Procurement and standard form contracts

Standard form contracts have always been a feature of the construction industry. A summary of the main standard forms relevant to commercial developments is set out below and is followed by a brief commentary on the effects of the trends.

Professional appointments

- **Architect – RIBA SFA/99** (updated April 2000).
- **Engineers – ACE conditions of engagement 2002.** A suite of conditions to accommodate particular engineering services, including also a short form agreement and sub-consultancy agreement.
- **Quantity/Building Surveyor – RICS Form with guidance notes 1999** for quantity surveying services and **RICS Form with guidance notes, 2nd edition 2000** for building surveying services.

Development and procurement

- **Planning Supervisor** – The **RIBA, RICS** and **ACE** standard form appointments make provision for planning supervisor services. It is common for the architect, quantity surveyor or engineer to also undertake the role of planning supervisor. Sometimes an independent consultant is appointed to undertake the planning supervisor role.
- **Project Manager** – **RICS Project Management Memorandum of Agreement and Conditions of Engagement 3rd Edition September 1999** and **RIBA Form of appointment for Project Manager PM99.**
- **General** – **The ICE Professional Services Contract, 2nd Edition 1998.** This is part of the New Engineering Contract.

Construction contracts

- **JCT 1998 Private with Quantities; Private without Quantities; and Private with Approximate Quantities** (the latest forms also incorporate amendments 1-3 or 1-4). The most commonly used traditional building contract forms. There are useful supplements to provide for sectional completion and specialist design, which may be incorporated. There are also local authority versions of these forms.
- **JCT 1998 With Contractor's Design** (with amendments 1-4). The JCT design and build form of contract.
- **JCT 1998 Management Contract** (with amendments 1-3). The employer contracts with the management contractor who in turn contracts with the trade contractors. The management contractor has limited liability. This form is now rarely used.
- **JCT Construction Management Contract January 2002.** The employer contracts directly with the trade contractors. The construction manager project manages under a separate contract with the employer.
- **The ICE New Engineering Contract.** This comprises the **Engineering and Construction Contract, 2nd Edition 1995** including options for a Priced Contract, a Target Cost Contract, a Cost Reimbursable Contract and a Management Contract.
- **The NEC Partnering Option X12, 2001.**
- **The ACA standard form contract for Project Partnering, PPC2000.**
- **The ACA standard form of Specialist Contract for Project Partnering, SPC2000.** This is a specialist subcontractor version of the PPC2000 form of contract. It can be used with PPC2000 or on its own.
- **Sub-Contracts** – standard forms of sub-contract commonly used are **Construction Confederation Dom/1** (being replaced by the new JCT DSC/C subcontract form) and **Dom/2**, the **JCT Standard Form of Domestic Sub-Contract 2002, DSC/C** and the **Engineering and Construction Subcontract, 2nd edition 1995.** The new JCT form of subcontract is arguably a more collaborative form with a better balance of risk between contractor and subcontractor. A 'with design' form is expected next year to replace Dom/2.

New developments in standard form contracts in line with the trends referred to above have seen the publication of the NEC Partnering Option and the ACA PPC2000. The NEC Partnering Option builds on the ECC suite of contract conditions, which were, at the time of their publication, a fresh approach to the drafting of contract conditions. The ECC has been adopted by a number of notable organisations with apparent success. It is fair to say that the ECC terms and conditions have yet to be fully tested and judicially interpreted, while there is extensive judicial guidance on the

Development and procurement

JCT forms of contract perhaps offering a greater level of ultimate certainty. Some might say, however, that too much judicial guidance is in itself a sign that the traditional construction procurement process and forms of contract are flawed. Only time will tell. With the addition of the NEC Partnering Option, the ICE has attempted to introduce a contractual document to incorporate a more integrated partnering ethos into the contractual rights and obligations of the parties. This does not, however, introduce a multi-party contract and still adopts the traditional two party approach.

The ACA has gone further with PPC2000. This adopts a radical approach as the first multi-party standard form contract. This is an interesting development and the form has been endorsed by a number of industry bodies. It remains to be seen, however, how the contract provisions and rights and liabilities between the various parties will be judicially interpreted in the event of a serious dispute.

Although the new forms of contract are worthy of serious consideration, it should be remembered that more traditional forms often work perfectly satisfactorily where proper preparation is undertaken and a positive and collaborative attitude is instilled in the team by the client from the outset. The key to this is sufficient front-end preparation:-

- ❖ The adoption of a suitable procurement structure for the needs of the project.
- ❖ Appropriate specific amendments to the base standard form contracts.
- ❖ The efficient management of design development.
- ❖ Supply chain management.

The contract administrator's role

It is a feature of most construction contracts that a person is appointed by the employer to administer the terms of the contract on the employer's behalf.

The contract administrator owes a duty of care to the employer. Under the terms of the contract he must undertake a number of administrative functions including the following:-

- ❖ Managing the client/contractor.
- ❖ Coordinating the pre-project, project and post project phases.
- ❖ Instigating client variations.
- ❖ Agreeing interim payments.
- ❖ Issuing certificates including payment certificates and practical completion.

The contract administrator has an important role in giving advice and information and also monitoring the work. However, he/she must also remain unbiased in matters such as certification of payments and ensuring that the contract terms are adhered to.

The mandatory nature of these duties is reflected in the contract between the contract administrator and the employer and in the contract between the employer and the contractor. As such, there is often considerable scope for disagreement between the contracting parties, both in contract and tort, on whether these duties have been satisfactorily performed.

It is with this background, that those fulfilling the contract administrator's role should be clear on what is required of them. The obligations set down by JCT 98 are detailed below.

Contract management

 Employer's agent

 The quantity surveyor's role

 Partnering

Development and procurement

Development and procurement

Employer's agent

Employer's agent duties

Increasingly, clients wish to procure projects in the pursuit of certainty of cost. This has led to the rise in the various forms of design and build procurement which transfer the 'risk' in any development to the contractor, while giving him greater flexibility to deliver the product. As an agent acting on behalf of the employer, it is essential that the following pre- and post-contract services are provided:

Pre-contract service

- ❖ Define the responsibilities of the employer, employer's agent and contractor
- ❖ Appraise and quantify the risks
- ❖ Formulate the employer's brief and identify specific requirements
- ❖ Assess the contractor's proposals and ensure compliance with the employer's requirements
- ❖ Undertake design audit of the contractor's proposals for compliance with the employer's requirements
- ❖ Evaluate the offer, the contract sum analysis and stage payments and assess value for money

Post-contract service

- ❖ Set up quality control procedure and report on works carried out on site
- ❖ Provide site visits and chair meetings
- ❖ Implement changes to the employer's requirements only on written approval of the client
- ❖ Agree stage payments and recommendations for payments
- ❖ Prepare monthly project control statements and cash flow forecasts to client
- ❖ Advise on practical completion, preparation of snagging schedules and component literature

General exclusions

- ❖ Checking and verifying contractor's design in terms of adequacy and efficiency
- ❖ Checking and verifying contractor's design in terms of fitness for purpose

In undertaking duties as the employer's agent, it is important to recognise the contractor's freedom to design, while respecting the client's brief and auditing the quality of the end product.

Development and procurement

The quantity surveyor's role

Capital project advice

The function/mission of the quantity surveyor may be defined as the optimisation of purchasing from the construction industry.

Several capabilities are required to fulfil the mission:-

- ❖ An ability to predict future costs from limited information and in dynamic market conditions.
- ❖ An ability to manage the procurement process to ensure that predictions of cost, time and quality are delivered.
- ❖ An awareness of risk with a capability to assess and manage that risk.
- ❖ An ability to demonstrate value for money.

Current trends in the industry lean towards cost reductions, with pressure coming from major purchasers; the importance of value for money is undoubtedly here to stay. Fixed out-turn costs are a necessity.

This is against a backdrop of upward cost pressure on suppliers of construction resources. Skills shortages are emerging and price rises are sticking.

The challenge to the construction industry and the quantity surveyor is sizeable. Changes in working practices and culture are inevitable and there needs to be a significant change in emphasis towards:

- ❖ value rather than cost;
- ❖ more professionalism in the early stages of project delivery;
- ❖ effective team work between the construction professions and contractors – adversarial relationships are no longer appropriate and will not survive;
- ❖ an understanding of the clients' business, and with that, the ongoing costs of the finished project;
- ❖ an awareness of financial incentives and opportunities available to the client (VAT relief and capital allowances – for details, see relevant sections in the cost management and taxation chapter); and
- ❖ an awareness on the part of purchasers that effective management of the construction process is a valuable service and does not come cheap.

Partnering

The traditional approach to construction procurement is driven by the terms of a contract. Relationships are imposed rather than developed; cost is focused at the expense of value; problems are packaged into liabilities rather than accepted as joint responsibilities; risk is off-loaded rather than managed. Partnering offers an alternative.

Partnering unites the sponsoring, design and construction teams and it endeavours to drive them forward with a common purpose.

The important characteristics are involvement, ownership and trust. This becomes obvious when one considers the facts:-

- ❖ The parties' objectives are not mutually exclusive.
- ❖ A construction project should not comprise a discrete stage of design followed by construction.

Development and procurement

- Generally, parties to a venture perform better if given a measure of control and a share in the action. This is all about ownership and being a stakeholder.

There are several levels of partnering. The simple indicators include the following:-

- Early involvement of contractors and subcontractors; viewing a construction team rather than a design team and a contractor.
- Flexibility – for example, being prepared to accept higher short-term costs in the interests of longer-term value; reducing cost rather than simply transferring it to others.
- A willingness to look beyond contractual responsibilities and provide more than is called for. 'The gold service' – going the extra mile.
- Concentration upon relationships rather than contractual positions. Actions driven by a common purpose rather than a book of rules.

More formal arrangements are implied in:

- project partnering; and
- strategic partnering.

The tools implied in partnering usually involve the following:-

- Two stage tendering – getting the contractor on board early.
- Negotiation – allowing the contractor to contribute to the design process rather than having to react to it; gaining a clear understanding of the project's objectives before committing to contract and ultimately being able to reduce his/her costs and, in that process, the client's costs. Perhaps more importantly, however, if the process of negotiation is conducted correctly the contractor is afforded the opportunity of understanding the client's definition of value.
- Open book – everything competitively tendered but with the leader in the procurement process, the contractor, fully acquainted with the project's decision-making rationale.
- A charter – as distinct from the rigidity of a contract, the parties set out what they want to achieve rather than what they are obliged to do.
- Performance related reward – allowing the contractor to become a stakeholder in the project budget rather than a conduit through which it flows.

Ultimately of course, partnering is all about attitude. The logic of partnering is inescapable; the extent to which it works will depend on the ability of the parties to shake free from traditional attitudes but also on the ability of the management team to deliver a project in unison and without (necessarily) conventional contract protection.

Development

 Construction management

 Project management

 Quality management and professional construction services

Development and procurement

Construction management

Under this form of procurement, the client appoints a construction manager who is paid a fee for managing and overseeing pre- and post-contract project activities. Site overheads may be included or paid direct by the client. Separate direct appointments will generally exist for the design team members. Building can begin as soon as the design is sufficiently advanced to allow the initial stages of construction to proceed and to prepare and agree a cost estimate for the entire scheme.

Construction is divided into works packages – either on single or multi-trade lines – and the client enters into direct contracts with each of the individual package contractors. This differs from management contracting where the contracts are between the management contractor and each package contractor. The individual packages are tendered at appropriate times throughout the construction period under the direction and management of the construction manager.

Whether a given project is suited to construction management (CM) will depend on a number of factors:

Size	CM is an involved form of procurement and can seldom be justified for small projects (less than £1 million certainly and generally much larger).
Complexity	CM is suited to complex projects with a substantial proportion of specialist package contractor involvement.
Uncertainty	Where the project is being conducted in an uncertain environment, CM affords clients a higher degree of flexibility to make changes during the process while minimising the consequent time and cost penalties.
Time	CM permits overlap between design and construction because tendering of packages can take place on a staggered basis rather than all at once as per traditional arrangements. This can save time and suit a 'fast track' approach. Greater involvement of the client with the CM and package contractors may also improve efficiency and introduce time savings. It should be noted however that time savings may be at the expense of cost risk as the scope isn't fixed until much later in the process.
Cost	Certainty of final cost is possible if the scope is finalised before the package contracts are placed. Obviously this sacrifices some of the flexibility referred to. This basis of contracting (in common with management contracting) should result in the lowest cost at the end of the day because the best price is chosen for each package contract without a main contractor's risk provision. Clients have more options to alter or adapt the design throughout the construction period without leaving themselves at a negotiating disadvantage with a main contractor. There is also greater potential to 'value engineer' the project under these arrangements.
Design	Often clients wish to retain control of the design process and this is facilitated under CM. Changes to design during construction carry less cost and time risk than under other procurement methods.

Development and procurement

Project management

Project management might be defined as the management of any activity that has a beginning and an end. Therefore, project management is directly relevant to all construction work and to many other industries. The application of specialised project management techniques in the construction industry has developed rapidly over the last 20 years or so. This is in response to the increasing complexity of projects and a perception that the industry is becoming fragmented into a large number of specialists which require active coordination.

The first task of a project manager is to define precisely the scope of the project that he or she is tasked with managing.

The next stage is to gain an understanding of the client's objectives in terms of time, cost and performance. These objectives need to be specified: completion by a particular date; expenditure not to exceed a certain amount; and creation of a particular building to a certain specification.

The relative importance of these objectives also needs to be ascertained. For instance, a client may have an overriding need to complete by a certain date and may be prepared to incur extra costs in order to do so. The project manager needs to understand this when allocating contingency sums and programme float periods.

The objectives then need to be conveyed to the design and construction team, as this knowledge will affect decision making throughout all stages of the project. The client's brief is a document produced by the project manager for this purpose.

An effective project manager must be skilled in leading, directing and coordinating a project team drawn together specifically for the duration of the project. This requires personal, presentational and above all communication techniques.

Control of time, cost and performance requires the establishment of master control documents, such as the project programme and cost plan. These allow the project manager to monitor progress, evaluate status and manage delivery. Other more detailed control documents are also required.

Successful project management entails the following:

- ❖ identification of time, cost and performance objectives in absolute and relative terms;
- ❖ communicating objectives to the team;
- ❖ establishment of strategic as well as detailed controls;
- ❖ acknowledgement that only the remaining time, cost and performance can be managed; and
- ❖ active management.

Quality management and professional construction services

Quality assurance in the form of BS EN ISO 9001:1994 has been with us for the last eight years and, following consultation with the users, was reissued as BS EN ISO 9001:2000 on 15 December 2000. This is intended to be a simpler, more flexible document for organisations to adopt and use.

On the BS EN ISO 9001:2000 publication, BS EN ISO 9002:1994 and BS EN ISO 9003:1994 were withdrawn as they have been incorporated within the new document. It should be noted that the revised title, 'Quality

Development and procurement

management systems – requirements,' reflects the change from assuring quality to producing a quality management system.

Most of the 'old' system is covered by the new document and the content has been revamped into five main sections instead of the original 20. The five sections are identified as follows:

Quality management system – this is a system for the control of documentation along with the Quality Manual and controls for looking after records.

Management responsibility – this requires management to set policies and objectives, to review the systems and to communicate to the organisation the effectiveness of the system.

Resources management – this covers not only people but also the physical resources such as equipment, premises and any support services required.

Production realisation – this deals with the process required to deliver to the customer the product/service they require having first identified what that need is.

Measurement analysis and improvement – this is the measurement and monitoring of the management system, customer satisfaction and the analysis of data for the continuing improvement of the system.

The implementation of BS EN ISO 9001:2000 is on the basis of the plan-do-check-act principle:

Plan

- ❖ Identify customer needs and expectations
- ❖ Strategic planning
- ❖ Set polices and objectives

Do

- ❖ Implement and operate the processes

Check

- ❖ Collect business results
- ❖ Monitor and measure the processes
- ❖ Review and analysis

Act

- ❖ Continually improve process performance

Transition

If you are considering accreditation to BS EN ISO 9001:2000, then your new quality management system will be based on the revised standards. If, however, you have an accredited system then you are encouraged to make the transition as soon as possible. This involves additional training of certified auditors/lead auditors and a period of assessment by your accredited body before registration can be obtained to BS EN ISO 9001:2000.

Building legislation and control

 Acts of parliament and regulations

 The Fire Precautions Act 1971

 Fire Precautions (Workplace) Regulations 1997 and (Amendment) Regulations 1999

 The Disability Discrimination Act 1995

Legislation and regulation

Legislation and regulation

Acts of parliament and regulations

England and Wales

There are some 170 national Acts and some 220 local Acts relating to the design and construction of buildings in England and Wales. Listed below are some of the more relevant.

- The Access to Neighbouring Land Act 1992
- The Agriculture (Miscellaneous) Act 1968
- The Ancient Monuments & Archaeological Areas Act 1979
- The Arbitration Act 1996
- The Betting, Gaming and Lotteries Act 1963
- The Betting, Gaming and Lotteries (Amendment) Act 1985
- The Building Act 1984
- The Building Regulations 2000
- The Building (Approved Inspectors) Regulations 2000
- The Building (Approved Inspectors etc) (Amendment) Regulations 2002
- The Building (Prescribed Fees) Regulations 1994
- The Building (Amendment) Regulations 2001
- The Building (Amendment) Regulations 2002
- The Capital Allowances Act 1990, 2001
- The Celluloid and Cinematograph Film Act 1922
- The Chronically Sick and Disabled Persons Act 1970
- The Chronically Sick and Disabled Persons (Amendment) Act 1976
- The Civil Procedure Rules 1998
- The Clean Air Act 1956, 1968 and 1993
- The Coal Mining Subsidence Act 1991
- The Commonhold and Leasehold Reform Act 2002
- The Construction (Design & Management) Regulations 1994, 2000
- The Construction (Health, Safety and Welfare) Regulations 1996
- The Contaminated Land (England) Regulations 2000
- The Contaminated Land (England) (Amendment) Regulations 2001
- The Contaminated Land (Wales) Regulations 2001
- The Contracts (Rights of Third Parties) Act 1999
- The Control of Asbestos at Work Regulations 1987
- The Control of Asbestos at Work (Amendment) Regulations 1998
- The Control of Asbestos at Work Regulations 2002
- The Control of Lead at Work Regulations 1998
- The Control of Lead at Work Regulations 2002
- The Control of Substances Hazardous to Health Regulations 1999
- The Control of Substances Hazardous to Health Regulations 2002
- The Control of Pesticides Regulations 1986
- The Control of Pollution Act 1974
- The Control of Pollution (Amendment) Act 1989
- The Countryside and Rights of Way Act 2000
- The Defective Premises Act 1972
- The Disability Discrimination Act 1995
- The Disabled Persons Act 1981
- The Education (School Premises) Regulations 1999
- The Electricity Act 1947, 1957, 1972, 1989
- The Electricity at Work Regulations 1989
- The Environmental Protection Act 1990
- The Environment Act 1995
- The Factories Act 1961
- The Fire Precautions Act 1971
- The Fire Precautions (Workplace) Regulations 1997

Legislation and regulation

The Fire Precautions (Workplace) (Amendment) Regulations 1999
The Fire Safety and Safety of Places of Sport Act 1987
The Food Safety Act 1990
The Gaming Act 1968
The Gaming (Amendment) Act 1980, 1982, 1986, 1987 and 1990
The Gas Safety (Installation & Use) Regulations 1998
The Health & Safety (Safety Signs & Signals) Regulations 1996
The Health & Safety at Work etc. Act 1974
The Health & Safety (Miscellaneous Amendments) Regulations 2002
The Highways Act 1980
The Highways (Amendment) Act 1986
The Historic Buildings and Ancient Monuments Act 1953
The Housing Act 1957, 1961, 1964, 1969, 1974, 1980, 1985, 1988 and 1996
The Housing and Building Control Act 1984
The Housing Defects Act 1984
The Housing Grants, Construction and Regeneration Act 1996
The Insolvency Act 2000
The Land Registration Act 2002
The Landlord and Tenant Act 1954, 1985, 1987, 1988
The Landlord and Tenant (Covenants) Act 1995
The Latent Damage Act 1986
The Late Payment of Commercial Debts (Interest) Act 1998
The Law of Property (Miscellaneous Provisions) Act 1989
The Leasehold Reform, Housing and Urban Development Act 1993
The Licensing Act 1964, 1988
The Licensing (Amendment) Act 1980, 1981 and 1985, 1989
The Local Government Act 1999, 2000
The Local Government and Housing Act 1989
The Local Government Finance Act 1992
The Local Government (Miscellaneous Provisions) Act 1976 and 1982
The Local Government Planning and Land Act 1980
The Local Government Act 1963, 1972, 1985 and 1992
The London Building Act 1930
The London Building (Amendment) Act 1935, 1939
The National Heritage Act 1980 and 1983
The National Heritage Act 2002
The National Lottery Act 1998
The New Roads & Street Works Act 1991
The Offices Shops and Railway Premises Act 1963
The Party Wall etc. Act 1996
The Petroleum (Consolidation) Act 1928
The Planning & Compensation Act 1991
The Planning (Consequential Provisions) Act 1990
The Planning (Hazardous Substances) Act 1990
The Planning (Listed Buildings and Conservation Areas) Act 1990
The Planning (Listed Buildings and Conservation Areas) (Scotland) Act 1997
The Private Places of Entertainment (Licensing) Act 1967
The Property Misdescriptions Act 1991
The Public Health Act 1936 and 1961
The Rights of Light Act 1959
The Safety of Sports Grounds Regulations 1987
The Scheme for Construction Contracts (England and Wales) Regulations 1998
The Special Educational Needs and Disability Act 2001

Legislation and regulation

The Theatres Act 1968
The Town and Country Planning Act 1990
The Town & Country Planning (Environmental Assessment & Permitted Development) Regulations 1995
The Town & Country Planning (Environmental Assessment) (England & Wales) Regulations 1999
The Water Act 1945, 1948, 1973, 1981, 1983 and 1989
The Warm Homes and Energy Conservation Act 2000
The Water Industry Act 1999
The Water Resources Act 1991
The Workplace (Health, Safety and Welfare) Regulations 1992

Northern Ireland

There are some 108 national Acts relating to the design and construction of buildings in Northern Ireland.

The Building (Prescribed Fees) Regulations (Northern Ireland) 1997
The Building Regulations (Northern Ireland) 2000
The Construction (Design & Management) Regulations (Northern Ireland) 1995
The Construction (Design & Management) (Amendment) Regulations (Northern Ireland) 2001
The Control of Asbestos at Work Regulations (Northern Ireland) 1988
The Control of Asbestos at Work (Amendment) Regulations (Northern Ireland) 2000
The Control of Lead at Work Regulations (Northern Ireland) 1998
The Control of Substances Hazardous to Health Regulations (Northern Ireland) 2000
The Defective Premises (Landlord's Liability) Act (Northern Ireland) 2001
The Factories Act (Northern Ireland) 1965
The Fire Precautions (Workplace) Regulations (Northern Ireland) 2001
The Gas Safety (Installation & Use) Regulations (Northern Ireland) 1997
The Health & Safety (Safety Signs & Signals) Regulations (Northern Ireland) 1996
The Office & Shop Premises Act (Northern Ireland) 1966
The Planning (Compensation, etc.) Act (Northern Ireland) 2001
The Planning (Hazardous Substances) (Northern Ireland) Act 1993
The Scheme for Construction Contracts (Northern Ireland) 1999
The Workplace (Health, Safety & Welfare) Regulations (Northern Ireland) 1993

Scotland

There are some 120 national and 50 Scottish Acts relating to the design and construction of buildings in Scotland.

NB: amending Acts (those shown with certain Acts) have important effects and should be read with them.

The Betting, Gaming and Lotteries Act 1963
The Building (Scotland) Act 1959 and 1970 as amended by the Housing (Scotland) Act 1986
The Building (Scotland) (Amendment) Regulations 1997
The Building Standards (Scotland) Regulations 1990 (inc. amendments 1993-99)
The Building Standards (Scotland) Amendment Regulations 2001
The Building Standards (Scotland) Amendment Regulations 2001 Amendment Regulations 2002
The Building Standards and Procedure Amendment (Scotland) Regulations 1999
The Celluloid and Cinematograph Film Act 1922

Legislation and regulation

The Chronically Sick and Disabled Persons Act 1970 as extended by the Chronically Sick and Disabled Persons (Scotland) Act 1972 as amended by the Chronically Sick and Disabled Persons (Amendment) Act 1976
The Cinemas Act 1985
The Civic Government (Scotland) Act 1982
The Clean Air Acts 1956, 1968 and 1993
The Contaminated Land (Scotland) Regulations 2000
The Control of Pollution Act 1974
The Control of Pollution (Amendment) Act 1989
The Electricity (Scotland) Act 1979
The Factories Act 1961
The Fire Precautions Act 1971
The Food Safety Act 1990
The Gaming Act 1968
The Gas Act 1972, 1986 and 1995
The Health & Safety at Work Act 1974 as amended by the Building Act 1984
The Housing (Scotland) Act 1987,1988, 2001
The Late Payment of Commercial Debts (Scotland) Regulations 2002
The Licensing (Scotland) Act 1976
The Licensing (Amendment) (Scotland) Act 1992
The Local Government etc. (Scotland) Act 1994
The Local Government and Planning (Scotland) Act 1982
The Local Government (Miscellaneous Provisions) (Scotland) Act 1981
The Local Government and Housing Act 1989
The Local Government (Scotland) Act 1973, 1975, 1978 and 1994
The National Heritage (Scotland) Act 1985
The Offices, Shops and Railway Premises Act 1963
The Planning & Compensation (Scotland) Act 1991
The Planning (Consequential Provisions) (Scotland) Act 1997
The Planning (Hazardous Substances) (Scotland) Act 1997
The Planning (Listed Buildings & Conservation Areas) (Scotland) Act 1997
The Pollution (Prevention & Control) Scotland Regulations 2000
The Roads (Scotland) Act 1984
The Safety of Sports Grounds Regulations 1987
The Scheme for Construction Contracts (Scotland) Regulations 1998
The Scotland Act 1998
The Sewerage (Scotland) Act 1968
The Theatres Act 1968
The Town and Country Planning (Scotland) Act 1997
The Water (Scotland) Act 1980
The Water Industry (Scotland) Act 2002

Statutory instruments and orders

Many Acts of parliament empower secretaries of state and ministers to publish Statutory Instruments and Orders to implement legislation which, although on the Statute Book, require 'enactment'. The Building Regulations are a prime example. In these cases guidance as to how acts ought to be interpreted is given.

The Fire Precautions Act 1971

Under the 1971 Act, specifically designated classes of premises are required to have a fire certificate issued by the Fire Authority. Currently this applies to:

Legislation and regulation

Hotels and boarding houses – except where no more than six people (staff and/or guests) have sleeping accommodation all on ground and/or first floors

Factories, offices, shops and railway premises – except where not more than 20 people work at any one time, or not more than ten persons work on any floor other than the ground floor, and shops and offices where only the employer's near relatives work or where not more than 21 hours weekly are worked.

In dealing with applications, the Fire Authority has to ensure that the means of escape from fire is adequate and effective and that the facilities for fighting fire and/or of giving warning in the event of fire achieve 'reasonable standards'. The existing or proposed use of the premises will be relevant.

If the standard is satisfactory a fire certificate must be issued but if it is not the Fire Authority must state the improvements required before they are prepared to issue a certificate.

A certificate will specify:

- ❖ the particular use or uses of the premises which the certificate covers;
- ❖ the means of escape in case of fire, including fire doors and fire escapes;
- ❖ the measures taken for securing safe and effective means of escape; and
- ❖ the type, number and location of the fire fighting equipment and fire alarms or warning systems in the premises.

A certificate can impose requirements ensuring that escape routes are kept clear, equipment is properly maintained and employees are adequately trained in fire drills. Further requirements may limit the number of people able to occupy the premises at any one time and can relate to other factors necessary to reduce the risk to persons in case of fire. The requirements can be imposed on the whole, or parts of the premises and may apply different requirements to the various sections. If the Fire Authority's requirements regarding the issue of a certificate seem unreasonable, the applicant may make appeal to a magistrate's court within 20 days of receipt.

Even if the premises are exempt under the act or have been granted exemption by the Fire Authority, the Fire Precautions Act imposes a duty on the occupier and / or owner to ensure that reasonable provision is made for means of escape and means of fighting fire. There are penalties for contravening this duty.

The effect of the Fire Precautions (Workplace) Regulations 1997 and (Amendment) Regulations 1999 must be remembered. These regulations are described more fully below, but essentially supplement the existing provisions of the Fire Precautions Act. The 1971 act still stands with the certification regime in force. However, there is now an additional responsibility placed on all employers to ensure that their premises have adequate fire precautions in place. This covers a wider range of premises than those covered in the 1971 Act, there now being only a very small number of workplaces that are exempt.

Fire Precautions (Workplace) Regulations 1997 and (Amendment) Regulations 1999

Following the introduction of the Fire Precautions (Workplace) Regulations on 1 December 1997 the Fire Precautions (Amendment) Regulations 1999

Legislation and regulation

have now been put in force from 1 December 1999. These are in response to criticism from the European Commission on the failure of the earlier regulations to comply with the original directives. They exempted many premises that should have been covered and failed to reflect fully the unconditional nature of employers' responsibilities, namely that it ultimately rests with the employer, or the person responsible for the building (for example, the landlord), to ensure compliance with the regulations.

Employer's duties

The 1997 regulations introduced fire to the realm of the Management of Health & Safety at Work Regulations 1992 (MHSW). These require employers to treat fire in the same way as any other health and safety issue and ensure that it is specifically covered in the risk assessment and in the comprehensive duties that follow the assessment. These include the provision of preventative protective measures, the use of competent persons, information provided to employees, incorporation and coordination with other employers and procedures for workers to follow where there is serious or imminent danger.

When the 1997 regulations were introduced it was the general assumption within the industry, and the Home Office's opinion, that those buildings covered by the UK system of fire certification would adequately comply. It has now been made clear that this is not the case and hence the new (amendment) regulations bring all work places under the regulations.

The general requirements are as follows:-

- ❖ Provide adequate fire fighting equipment, fire detectors and alarms. Any non-automatic fire fighting equipment such as extinguishers must be accessible, simple to use and indicated by signs. Employers must also take measures for fire fighting: nominating employees to implement those measures; ensuring they are trained and fully equipped; and making any necessary contact with the emergency services.
- ❖ Keep emergency routes and exits clear. Routes should lead to a place of safety as directly as possible, be indicated by signs and provided with emergency lighting. It must also be possible to evacuate the workplace as quickly and safely as possible, with emergency doors opening in the direction of escape.
- ❖ Maintain the equipment, routes and exits in sufficient state, working order and good repair.

It is important to stress that having a current fire certificate in place is not regarded as adequate demonstration of complying with the regulations. The emphasis has been very heavily placed upon the employer or the landlord, not only to ensure, but be able to demonstrate, that the appropriate risk assessments have been undertaken and appropriate action taken upon the findings of those risk assessments.

Enforcement of these regulations will still be carried out by the Fire Authority which will have the power to issue notices of non-compliance stipulating those works that must be undertaken within a set period of time. Failure to comply with the regulations or the requirements, or an enforcement notice will be a criminal offence.

The Fire Safety and Safety of Places of Sport Act 1987

This Act has made minor technical amendments to the FPA in the areas of fire certificate exemption; charges for the certification works; means of escape and fire fighting; interim duties as to the safety of premises; premises involving serious risk to persons; inspection of premises and other miscellaneous minor changes.

The remainder of this Act concerns general safety with amendments to the Safety of Sports Grounds Act 1975, primarily extending the scope of the

Legislation and regulation

Act with similar provisions for the safety of stands at sports grounds. The Act also includes fire schedules relating to the above.

The Disability Discrimination Act 1995

The purpose of the Disability Discrimination Act is to prevent discrimination against disabled people. Disabled people should not be treated less favourably in connection with employment, the provision of goods, facilities or services to the public or by those selling, letting or managing premises.

The Act is split into different sections.

Part I - defines the term 'disability'
Part II - deals with discrimination in employment
Part III - deals with discrimination in the provision of goods, facilities and services and the disposal and management of property
Part IV - deals with provisions for education
Part V - deals with provisions for public transport
Part VI - deals with the setting up of the National Disability Council
Part VII - supplemental provisions
Part VIII - miscellaneous provisions

Part I - Definition of disability within the Act

A physical or mental impairment, which has a substantial and long-term (12 months minimum) adverse effect on a person's ability to carry out normal day to day activities.

Part II - Employment

This section of the Act came into force on 2 December 1996 and comes into effect when a disabled person is employed, or an employee becomes disabled. The Act places a duty on employers of 15 people or more to provide accessible facilities for all disabled employees and to take all reasonable measures to enable disabled employees to carry out their work. An employer is expected to take reasonable measures to allow a person to do their job. This may involve:

- making adjustments to premises;
- moving a disabled person's place of work;
- altering hours of work;
- reallocating a disabled person's duties;
- acquiring or modifying equipment; and
- providing a reader or interpreter.

Part III - The provision of goods, facilities and services including discrimination in relation to premises

What is covered by this part of the Act?

- Any place where the public may enter.
- Accommodation in hotels, boarding houses' etc.
- All retailers or tradesman's premises.
- Any building owned by a public authority.
- Facilities for entertainment or recreation.

Legislation and regulation

Discrimination is deemed to have arisen if a provider of services treats a disabled person less favourably than he/she would treat others and:

- he/she cannot show the treatment in question was justified; and
- he/she failed to undertake his/her duty to make adjustments as described below.

Part III has and will come into force in three distinct sections.

The following duties came into force in December 1996

- A duty not to discriminate.
- A duty not to refuse service.
- A duty not to provide a worse standard of service.
- A duty not to offer worse terms.

The following duties came into force in October 1999

- A duty to change practices, policies and procedures that are discriminatory.
- A duty to provide extra help such as auxiliary aids.
- A duty to overcome physical restrictions by the provision of alternative methods.

The following duties come into force in October 2004

Where a physical barrier makes use of any service which is offered to the public impossible or unreasonably difficult, a service provider must take reasonable steps to:

- remove the feature; or
- alter it so it no longer has an effect; or
- provide a reasonable means of avoiding the feature; or
- provide a reasonable alternative method of making the service available to disabled people (in force since 1 October 1999).

The Act does not specify what is reasonable, as it varies according to the characteristics of each area of provision. Account may be taken of:

- the type of services being provided;
- the nature of the service provider and its size and resources;
- the effect of the disability on the individual disabled person;
- the amount of time that the service provider has had prior to that date to make preparations;
- whether taking any particular steps would be effective in overcoming the difficulty that disabled people face in accessing the services in question;
- the extent to which it is practicable for the service provider to take the steps;
- the financial and other costs of making the adjustment;
- the extent of any disruption which taking the steps would cause;

Legislation and regulation

❖ the amount of any resources already spent on making adjustments; and
❖ the availability of financial or other assistance.

The Act does not require a service provider to take any steps which would fundamentally alter the nature of its service, trade profession or business.

Disability Discrimination (Providers of Services)(Adjustment of Premises) Regulations 2001

Applies to service providers and landlords of premises occupied by service holders.

The regulations prescribe particular circumstances in which it is reasonable or unreasonable for service providers to make physical alterations to the premises they rent or own or for lessors to withhold their approval to same.

Part IV - Education

Schools

This part of the Act builds on the Education Act 1993 which aims to provide all pupils with special educational needs, including disabled pupils, with an education and school place appropriate to their requirements.

Education authorities have a duty to place such children in mainstream schools subject to the wishes of their parents, and if the placement:

❖ is appropriate to the child's needs;
❖ does not conflict with the interests of other children in the school; and
❖ is an efficient use of the local educational authorities' resources.

Further education

This part of the Act builds upon the Further and Higher Education Acts of 1992, which require further education providers to take account of the needs of students with learning and other disabilities, when fulfilling duties to provide education.

The Act provides an additional duty on Further Education Funding Councils in England and Wales to:

❖ publish disability statements; and
❖ report to the government on their progress and their future plans for providing further education to students with disabilities.

(See also Special Educational Needs and Disability Act 2001)

Part V - Public transport

The Act enables the government to make regulations, known as accessibility regulations, to require that all new public transport vehicles – buses, coaches, trains and trams – are accessible. Accessibility regulations can also be made to cover newly licensed taxis. These regulations ensure disabled people can:

❖ get on and off public transport vehicles in safety; and
❖ travel in them in safety and reasonable comfort.

Part VI - The National Disability Council

The Act sets up two new independent statutory bodies to advise the government about disability issues and on the implementation of the Act.

Legislation and regulation

These bodies are:

- ❖ The National Disability Council in England, Wales and Scotland; and
- ❖ The Northern Ireland Disability Council.

The councils will advise the government about eliminating and reducing discrimination against disabled people and the operation of the act.

Codes of practice and advice notes

The government has drawn up various codes of practice to help implement the Act and is currently considering two more, in respect of educational establishments.

Further sources of information

Further information can be found on the website of the Disability Rights Commission at www.drc-gb.org.

Disability and The Building Regulations

The government is (at November 2002) conducting a second round of consultation on proposed amendments to Part M of the Building Regulations and an approved document relating to access and facilities for disabled people. No changes are proposed to the regulations as they relate solely to dwellings, but they will apply to non-domestic and mixed use buildings. Sections 1-5 of the draft approved document have been re-ordered to reflect the recently re-published BS 8300.

The objective of the proposed amendments is to ensure that new buildings meet reasonable standards of accessibility and to secure cost-effective improvements to the accessibility of existing building stock when other intended work is carried out. However, in the case of alterations, it is proposed that provided there is a suitable means of access to the alteration, there is no obligation to carry out improvements within the remainder of the building.

All exemptions must be justified through the use of "Access Statements".

The proposed Part M extends beyond meeting the access requirements of disabled people and to 'parents with children' and 'elderly people'. Clearly, providing access for children to an industrial unit may, for example, be unreasonable, and so the access statement would indicate the reasons for non-compliance. In reality, the provisions for children or the elderly probably do not extend the general duty that far anyway – most provisions for disabled people would serve all building users, in other words it is difficult to envisage any particular amendments that would not already cater for the young or elderly.

Legislation and regulation

Building and construction regulations

- ▶ The Building (Amendment) Regulations 2001

- ▶ Airtightness

- ▶ The Provision of Part II Housing Grants, Construction and Regeneration Act 1996

- ▶ Health and safety at work

- ▶ Construction (Health, Safety and Welfare) Regulations 1996

- ▶ The Construction (Design and Management) Regulations 1994

Legislation and regulation

The Building (Amendment) Regulations 2001

The 1 April 2002 saw the imposition of Parts L1 and L2 to the Building Regulations covering the conservation of fuel and power in dwellings and buildings other than dwellings respectively.

The driving force behind the gradual imposition of tighter regulations is the recognition that buildings are responsible for some 50% of carbon dioxide emissions. For example in a typical UK home, heating accounts for 57% of all energy consumed and produces over one tonne of carbon dioxide annually. The government has a manifesto aim of reducing emissions to below 1990 levels by 2010, and has signed up to the Kyoto protocol which requires EU countries to cut greenhouse gas emissions by 8%. Without doubt, this means that thermal standards will become even more onerous in the future.

Part L2 requires that reasonable provision shall be made for the conservation of fuel and power in buildings or parts of buildings other than dwellings by:

- ❖ limiting heat losses and gains through the fabric of a building;
- ❖ limiting heat loss; and
- ❖ limiting exposure to overheating.

(Other requirements relate to energy efficiency of lighting, heating and air conditioning).

The Approved Documents indicate that the requirements can be satisfied by:

- ❖ limiting the heat loss through the roof, wall, floor, windows and doors etc by suitable means of insulation, and where appropriate permit the benefits of solar heat gains and more efficient heating systems to be taken into account;
- ❖ limiting the heat gains in summer; and
- ❖ limiting heat losses (and gains where relevant) through unnecessary air infiltration by providing fabric which is reasonably airtight.

On the basis that heat loss through the glazed elements of a building is in the order of 40% of the total, a small improvement in thermal performance will have a big impact on the total heat loss. The expectation is that U values for glazed elements will be further tightened, from the (April 2002) level of 2.2 W/m^{2}K down to 2.0 W/m^{2}K probably in 2004 and 1.8 W/m^{2}K by the end of the decade

The Scottish Regulations (Part J), as published, are stricter than those applicable in England and Wales and they relate to the efficiency of the domestic heating systems.

Historic buildings and replacement windows

The need to conserve the special characteristics of buildings needs to be recognised. Work to these buildings should aim to improve energy efficiency without prejudicing the character of the building or affecting its long-term durability. (see SPAB information sheet No 4 'The Need For Buildings To Breathe'). The approved documents indicate that in order to achieve a suitable balance, the advice of the local authority's conservation officer should be sought. (see also BS 7913 'The Principles of the conservation of historic buildings').

The scope of 'historic buildings' has been widely defined. It includes:

- ❖ listed buildings;
- ❖ buildings situated in conservation areas;

Legislation and regulation

- buildings of local architectural and historical interest and which are referred to as a material consideration in a local authority's development plan; and
- buildings within national parks, areas of outstanding natural beauty, and world heritage sites.

The regulations state: "In arriving at an appropriate balance between historic building conservation and energy conservation, it would be appropriate to include restoring the historic character of a building that had been subject to previous inappropriate alteration, for example, replacement windows".

Replacement windows must also make performance elements covered in other approved documents no worse than the original products, including:

- means of escape (Part B);
- provisions for room ventilation (Parts F and J);
- means of access (Part M); and
- safety glazing (Part N).

Under the regulations

- Replacement windows and doors will be covered for the first time.
- Replacement windows can be treated either as for new build or must use glass with a centre pane U-value of 1.2 W/m^{2}K or less.
- When using replacement windows and doors, consideration must also be given to compliance with other parts of the Building Regulations notably parts B, E, F, J and N.
- Replacement doors will only need to meet the requirement where they have 50% or more glass area (measured overall of frame). Then they must be proved by hot-box testing, calculation or by using glass with a centre pane U-value of 1.2 W/m^{2}K.

Airtightness

Parts L1 and L2 of the Building Regulations contain, among other things, a requirement for minimising air leakage from buildings.

Achieving an airtight building means following three essential steps:-

- Design for airtightness.
- Build for airtightness.
- Test for airtightness.

It is not practicable to construct a building and then try to make it airtight. Remedial sealing can be difficult and costly. By designing in airtightness at the drawing stage you can deal with air barrier continuity and sealing details at critical elements – and ensure long-term performance by specifying the correct seal or sealant.

The main air leakage problems in buildings occur typically:

- around doors, windows, panels and cladding details;
- in gaps where the structure penetrates the construction envelope;
- in service entries: pipes, ducts flues, ventilators;
- in porous construction: bricks, blocks, mortar joints; and
- in joist connections within intermediate floors.

Legislation and regulation

Designers should identify all the problem areas, for example, sealing around pipe entries, and spell out responsibility for finishing off in the contract documents.

Constructing the building to the airtightness specification is then down to the main contractor and sub-contractors. For this to be successful all of the work force should be aware of airtightness issues in the same way safety issues and codes of conduct are dealt with.

Inspection during construction is essential. Talking to and working with contractors is the best way of ensuring that the team understands the importance of the airtightness layer and how to incorporate it.

The only real way to be confident that the building meets an airtightness specification is to carry out a fan pressurisation test prior to handover as required in the proposed revisions to Part L. Large buildings, for example, hypermarkets or industrial buildings, need specialist larger capacity equipment.

If the three essential steps listed above are followed, the building should pass the test. In the event they are not and the building fails, the proposals in Part L state: "If on first testing the building fails to comply, the major sources of air leakage should be identified using the techniques described in TM23 [CIBSE Technical Memorandum]." This usually requires specialist help.

Despite improved understanding of construction techniques, few buildings are sufficiently airtight – this is true of new buildings as well as old. In a recent survey, only one out of 39 buildings tested met a good practice benchmark for airtightness. While the degree of leakage varies considerably from building to building, it is not unusual for the problem to be equivalent to having a 9m^2 hole in the building envelope.

Both the government and CIBSE regard airtightness as a serious issue and encourage protective measures.

Airtightness testing

CIBSE has produced a guide titled 'Testing Buildings for Air Leakage' (TM23: 2000)

The technique that is most commonly used to measure air leakage is 'fan pressurisation testing'. This involves a specially designed system of fan units that blow air into the building, and the measurement of air leakage from the building at various air pressures.

Testing can be carried out on any building from a large hypermarket, multi-storey office block or factory to a small store, office or even an individual room within a building. Testing a large office or superstore takes about three hours and is generally done out of business hours when the premises are closed.

The Provisions of Part II Housing Grants, Construction and Regeneration Act 1996

The following aims to set out a brief outline of the issues which need to be considered when determining whether contractual terms are compliant with Part II of the Housing Grants, Construction and Regulation Act 1996 or whether certain provisions will be incorporated into the contract by the Scheme for Construction Contracts.

Parties to a construction contract are free to negotiate and agree the terms and conditions under which the works and services are to be carried out. However, there are many instances where the contract fails to address or

Legislation and regulation

comply with the minimum requirements relating to adjudication and payment laid down by the Act. Consequently, certain provisions will automatically be incorporated into the contract by the Scheme for Construction Contracts.

When does the Act apply?
(S104 - 107)

The Act applies to:-

- ❖ Contracts entered into on or after **1 May 1998**.
- ❖ **Contracts in writing.** It is sufficient if the contract is evidenced in writing by an exchange of letters or by reference to a standard contract.
- ❖ Contracts for **construction operations** which include the construction, alteration, repair, maintenance, decoration, demolition and installation in buildings forming or to form act of the land and also architectural, design, surveying or engineering advice.
- ❖ The carrying out of construction operations in England, Wales or Scotland whatever the applicable law of the contract.

The Act does not apply to:-

- ❖ **Variations or instructions** issued after 1 May 1998 which relate to a contract entered into before that date.
- ❖ Contracts with **residential occupiers** for work on their property where they intend to occupy the property as their residence.
- ❖ Certain mining, drilling and extraction operations.
- ❖ Installation or demolition of plant or machinery or steelwork to support or provide access to plant or machinery on a site where the primary activity is nuclear processing, power generation, water or effluent treatment, or the production processing of chemicals pharmaceuticals, oil, gas, steel, food or drink.
- ❖ Manufacture and delivery of materials not involving installation.
- ❖ Artistic works.

Letters of intent may be subject to the Act where they are sufficient to amount to a legally binding contract in their own right.

Adjudication (S108)

The contract must:-

- ❖ Allow either party to refer a dispute to adjudication at any time.
- ❖ Provide for the appointment of an adjudicator within seven days.
- ❖ Require the adjudicator to reach his decision within 28 days **after** the dispute has been referred to the adjudicator (or a longer period if agreed by the parties).
- ❖ Allow the adjudicator and the party who referred the dispute to him to extend the period for his decision by up to 14 days.
- ❖ Impose a duty on the adjudicator to act impartially.
- ❖ Enable the adjudicator to take the initiative in ascertaining the facts and the law surrounding the dispute.

Legislation and regulation

- ❖ Provide for the decision of the adjudicator to be binding on the parties until the dispute is taken to arbitration or the Courts.
- ❖ Provide that the adjudicator is not liable for anything he does unless he acts in bad faith.

If the contract does not comply with all eight elements in full, all the purported adjudication provisions of the contract will be set aside and the adjudication procedures under the Scheme for Construction Contracts will apply. The procedures under the scheme cover the eight points listed above and introduce time limits.

Payment

The Act and the Scheme for Construction Contracts provide a 'menu' of payment provisions covering:

- ❖ payment by instalments;
- ❖ final payment;
- ❖ withholding payment; and
- ❖ conditional payment.

If a contract fails to comply with any one of the provisions from the 'menu' the relevant provisions from the scheme will apply. The remainder of the contractual provisions that do comply with the Act will remain intact. It is therefore possible to end up with a contract where the payment provisions are a mixture of express terms agreed between the parties and implied terms from the scheme.

Payment by instalments (S109 - 110)

A party to a construction contract is entitled to payment by instalments, stage payments or other periodic payments unless:-

- ❖ The contract specifies that the duration of the work is less than 45 days; or
- ❖ The parties agree that the work is estimated to take less than 45 days.

Where a contract falls within the 45 day limit, the right to instalment payments is excluded but all other payment provisions (notice of withholding payment, set-off, etc) will apply as will the adjudication provisions outlined above.

Where a contract falls within the 45 day limit, payment of the contract price falls due 30 days after completion of the work (or 30 days after the contractor's claim if later) and payment must be made within 17 days.

In all other cases, the parties can agree between themselves:

- ❖ the amounts of each payment;
- ❖ the intervals between each payment;
- ❖ the date each payment becomes due; and
- ❖ the final date by which each payment must be made.

If the contract does not contain a clear mechanism for determining each of these four elements they will be determined by the Scheme for Construction Contracts, namely:-

- ❖ The amount of each payment will be based on the value of the work and other costs to which the contractor is entitled during the payment interval.
- ❖ There will be 28 day payment cycles.
- ❖ Payment is due seven days after each 28 day period (or seven days after the contractor's claim for payment if later).

Legislation and regulation

- ❖ The final date for each payment is 24 days after each 28 day period (or 24 days after the contractor's claim for payment if later).

The contract must provide for the paying party to give notice within five days of the date on which each instalment becomes due, specifying the amount proposed to be paid and the basis on which it is calculated. Any attempt in the contract to vary or exclude this requirement will be ineffective and this provision will be implied by the Scheme for Construction Contracts.

Final payment

The contract must contain a clear mechanism for determining when the final payment due under the contract becomes payable and the final date by which that payment must be made.

The parties are free to agree the dates or periods within which the final payment is due and is payable but if there is no such mechanism, in accordance with the Scheme for Construction Contracts the final payment:

- ❖ is due 30 days after completion of the work (or 30 days after the contractor's claim for payment if later); and
- ❖ the final date for making the final payment is 47 days after completion of the work (or 47 days after the contractor's claim for payment if later).

Withholding payment (S111)

- ❖ No payment can be withheld unless a 'notice of intention to withhold payment' has been given, specifying the amount to be withheld and the grounds for withholding payment.
- ❖ The notice must be given before the final date for payment.
- ❖ The contract can specify how long before the final date of payment, the notice must be given (even if it is just one day) but if this notice period is not specified, the Scheme for Construction Contracts applies and at least seven days notice must be given.

Conditional payment (S113)

Any provision in a contract which makes payment conditional upon the paying party receiving payment from someone else is ineffective and the payment provisions of the scheme outlined above will apply.

The only exception is where the contract provides that payment may be withheld if the reason for non-payment is the insolvency of someone else in the payment chain.

Suspending performance (S112)

If any payment is not received by the final date for payment and a notice of withholding payment has not been served, the contractor may suspend work after giving seven days notice of his intentions.

Any contractual term purporting to extend this seven day period is probably ineffective.

The right to suspend performance ceases when the relevant payment is received.

The period in which to complete the works is automatically extended by the number of days of the suspension. Any attempt to exclude this right of extension is probably ineffective.

Legislation and regulation

There are no provisions under the Scheme for Construction Contracts dealing with loss and/or expense as a result of a suspension.

A dispute over the amount of any payment withheld does not give rise to the right to suspend where an appropriate notice to withhold has been given. In this situation the parties may resort to adjudication.

This gives a brief summary of the provisions of Part II of the Housing Grants, Construction and Regeneration Act 1996. It is not intended to be a detailed explanation of the provisions of the Act and we recommend that legal advice is sought on any specific issues.

Health and safety at work

Essential legal duties

The Health & Safety at Work etc. Act 1974 requires all chartered surveyors to ensure, so far as is reasonably practicable, the health and safety of themselves and any other people who may be affected by their work.

In essence, the places where surveyors' work must be safe and working practices must be clearly defined, organised, and followed to avoid danger. This requires safety training and the distribution of relevant information, followed up by diligent and regular supervision.

These duties extend to anyone who uses surveyors' professional services.

Those who lease part of their premises to other businesses also may be responsible for them with regard to safety matters.

In addition, all working people whether employees, managers, partners or directors (self-employed or not) must behave in a way that does not endanger themselves or others.

Health and safety policy statement

Every firm of surveyors which employs five or more people is obliged by the Act to draw up a health and safety policy statement, which should be kept up to date, with any significant revisions being notified to employees.

The Health & Safety Executive (HSE) in "Writing your health & safety policy statement: how to prepare a safety statement for smaller businesses" describes what the document should say and gives a useful pro-forma.

It is suggested that employees, especially when taking a new job, should satisfy themselves that they have seen and understood the company's policy statement and have been made familiar with safe working practices. If they consider that the Health & Safety at Work Act is not being followed, they ought to say so, and if necessary ask advice from their local HSE office (listed in the telephone directory).

Practical procedures

Chartered surveyors should identify the hazards they may encounter in practice, carry out a risk assessment and plan accordingly. The way they proceed will depend upon the working environment: when surveying old and derelict buildings, for example, there may be holes in floors, parts of the structure may be unstable or there may be health hazards. Particular care should be taken to protect against personal attack.

Control of Substances Hazardous to Health Regulations 1999

Under these regulations, employers are required to evaluate the risk of all products used that may be harmful to the health of their employees and take appropriate measures to prevent or control exposure.

Legislation and regulation

Workplace (Health, Safety and Welfare) Regulations 1992

These apply to all workplaces. The regulations set out the minimum requirements in respect of the provision and maintenance of the environmental conditions, space allocation, sanitary and welfare provision of employees. In addition, they impose particular safety requirements on forms of construction or circumstances which are considered to be high risk. They do not apply to construction sites.

Provision and Use of Work Equipment Regulations 1992

These apply to all workplaces. The regulations set out the minimum health and safety requirements for both new and second-hand equipment, including its suitability, maintenance and training in its correct use. They identify specific hazards that the employer must prevent or adequately control.

Manual Handling Operations Regulations 1992

These require the employer to try to avoid the need for employees to undertake any manual handling operations at work that involve a risk of their being injured. Where this is not reasonably practicable the risk must be assessed and suitable provision made, including equipment, instruction and training for safe manual handling.

Management of Health & Safety at Work Regulations 1999

These regulations are of a wide-ranging general nature and overlap with many others. They require the employer to carry out an assessment of the risks of the hazards to which his/her employees are exposed at work or to others arising from or in connection with this work. He/she must instigate appropriate protective or preventive measures, reviewing and amending these as necessary.

The employer must appoint a 'competent person' to provide assistance in respect of these duties. Emergency procedures must be put in force to deal with any serious and imminent danger. Employees must be informed of these measures and suitably trained where required. They are obliged to comply with these instructions and warn of any situation considered to be a serious and immediate danger to health and safety.

Personal Protective Equipment at Work Regulations 1992

Under these regulations the employer has a duty to provide and maintain suitable personal protective equipment including adequate instruction and training on its correct use when risks to health and safety cannot be avoided by other means. Employees have a duty to make full and proper use of such equipment provided and to report any loss or obvious defects.

Construction (Health, Safety and Welfare) Regulations 1996

The CHSW Regulations came into force on 2 September 1996, to form a single set of regulations applicable to construction work and construction sites. They consolidate, modernise and simplify much of the previous

Legislation and regulation

legislation and complete the EC directive on construction.

Their aim is to protect the health, safety and welfare of everyone "carrying out construction work" and also to protect those affected by the work.

The regulations impose requirements with respect to:

- the provision of safe places of work and safe access and egress thereto (regulation 5);
- the provision of suitable equipment to prevent falls (regulation 6);
- working on or near fragile material (regulation 7);
- the prevention of injury from falling objects (regulation 8);
- the stability of structures (regulation 9);
- the carrying out and supervision of demolition and dismantling and the use of explosives (regulation 10, 11);
- the safety of excavations, cofferdams and caissons (regulation 12, 13);
- the prevention of drowning (regulation 14);
- the movement of pedestrian and vehicular traffic (regulation15);
- the construction of doors, gates and hatches (regulation 16);
- the use of vehicles (regulation 17);
- the risks from fire, the provision of emergency routes and exits, the preparation and implementation of evacuation procedures and the provision of fire-fighting equipment, fire detectors and alarms (regulations 18, 19, 20, 21);
- the provision of sanitary and washing facilities, a supply of drinking water, rest facilities and facilities to change and store clothing (regulation 22);
- the provision of adequate fresh air, reasonable temperature and weather protection (regulations 23, 24);
- the provision of lighting, including emergency lighting (regulation 25);
- the marking and good order of a construction site (regulation 26);
- the safety and maintenance of plant and equipment (regulation 27); and
- training and supervision (regulation 28).

Points of note (including changes of current practice)

Application of the regulations

The Health and Safety Executive is, in certain circumstances, able to issue exemption certificates subject to conditions or time limitations deemed appropriate and which may be subsequently revoked if considered necessary.

Subjective requirements

As with other recent health and safety legislation, the regulations have moved away from prescriptive demands and most of the requirements are, to the extent that they are 'suitable and sufficient', to be determined on the basis of a risk assessment of the particular circumstances. In addition, many are qualified by their 'reasonable practicability', which means that the action to be taken should be proportionate to the risk involved.

Scaffolding

This is one instance where the requirements are prescriptive and the

Legislation and regulation

measurements stated mean that virtually all scaffolding and working platforms will require an intermediate guard rail or other means of physical protection, such as a brick guard.

Welfare

The responsibility for the provision of suitable and sufficient welfare facilities is now that of the "person in control of a construction site" as well as the employer of the site operatives or the self employed.

The washing facilities are to be appropriate to the nature of the work and there is an inference that showers may be required in certain circumstances.

The Health and Safety Executive is paying great attention to the adequacy of welfare facilities on sites and has pointed out that some wash hand basins will need to be of sufficient depth to enable the complete forearm to be immersed for proper cleaning. On occasion, basins have been kept deliberately shallow to minimise the water required.

Falls

Particular emphasis is put on the measures to be taken to prevent persons falling from a height and the regulations set out a hierarchy of alternative methods by which this may be achieved. This takes into account practicability, physical constraints and the duration of the work. The use of safety harnesses, as with all personal protective equipment, should be a last resort.

Prevention and control of emergencies

There are specific requirements to plan for unforeseen circumstances that may arise on construction sites. To prevent risk from fire, explosion, flooding and asphyxiation, procedures for evacuating the site must be established, including emergency escape routes and, where necessary, fire fighting equipment, fire detectors and alarm systems must be provided.

Traffic movement

There are requirements to ensure the safe use and movement of vehicles used in connection with construction work and the provision of safe traffic routes generally.

Further information

For further details, reference should be made to the regulations themselves (ISBN 0-11-035904-6 available from HMSO).

In addition, the HSE has produced a guidance publication 'Health & Safety in Construction' which describes practical ways of complying with the regulations. (ISBN 0-7176-1143-4).

The Construction (Design and Management) Regulations 1994

The Construction (Design and Management) Regulations (CDM) 1994 provide the framework for managing health and safety during the construction, repair, maintenance and demolition of civil engineering and building works.

They impose statutory duties on clients, designers and planning supervisors (a new role) and contractors.

The principal objectives are to:

❖ ensure proper consideration and coordination of health and safety issues at every phase of a project, from feasibility study to demolition;

Legislation and regulation

- obtain adequate allocation of resources (including sufficient time) to enable duties imposed by the regulations to be met;
- involve directly, allocate and share responsibilities between all participants (including the client); and
- promote the appointment of competent (from a health and safety viewpoint) designers, planning supervisors and contractors.

To achieve its aims, for each new project to which the regulations apply, two documents are required:-

Health and safety plan – in two stages, planning and construction, to convey health and safety information to the contractor during the tender and construction phases of the project.

Health and safety file – a record of information relevant to health and safety to be retained by the client, made available and used throughout the life of the building or structure to assist its safe maintenance, alteration or demolition.

The client must appoint a health and safety coordinator. This role is separated for the pre-contract planning and post-contract implementation phases of the project: ie the planning supervisor, who is ultimately responsible for the preparation and coordination of the safety plan, up to the appointment of the principal contractor. The principal contractor takes over responsibility for developing and implementing the safety plan during the construction phase of the project.

Construction Regulations 1994
Revised Approved Code of Practice 'Managing Health & Safety in Construction' published by Health & Safety Commission (HSG 224)

The new ACOP came into effect on 1 February 2002.

Its aim is to rectify confusion and shortcomings identified in the previous publication, in particular to improve the focus of work carried out by the various dutyholders and reduce bureaucracy.

The ACOP has been re-formatted in a more logical, user-friendly layout and emphasises the financial and other benefits of the CDM regulations, provided they are implemented appropriately.

It clarifies a number of matters which were previously often misinterpreted, by listing those matters which are not required to be included in health and safety documents or which are excluded from the responsibility of particular dutyholders, by giving examples of both good and bad practice and by changing the emphasis of particular issues.

The topics where there are significant changes are summarised below and, for each, the concern to be addressed is described (in italicised text) followed by the suggested means of improvement.

Competence and resources

Expense and wasted effort arising from use of assessment questionnaires and processes that are often excessive, unnecessary and inappropriate.

Assessments should: form part of the more general checks routinely carried out (for example, quality, finance and viability), focus on the particular project and be proportionate to the risks, size and complexity of the work.

Time and programme

Dutyholders not appointed early enough in the project, design considerations made too late and (design and construction) programmes unrealistic.

Legislation and regulation

- Planning supervisor, principal contractor and other key people must be appointed as soon as possible.
- Design risks to be identified and removed or reduced early in the design process, at the concept and scheme stages, not left until detail design.

Design and designers

Using unnecessarily sophisticated risk analysis techniques; identifying risks but not adapting the design to remove or reduce them and overlooking less obvious 'designers'.

- A team approach to risk consideration is recommended.
- ACOP lists untypical designers, including temporary works engineers, interior designers, shopfitting, trade contractors and manufacturers of purpose-made products.
- ACOP lists, ways in which the designer can have direct positive influence on risks and circumstances in which relevant information must be provided to others together with those matters which are not the designer's responsibility.

Clients

Failing to appreciate the full scope of the definition of a 'client'; unnecessarily monitoring performance of others and not providing information, either at all or in sufficient time.

- ACOP gives examples of other parties who may be 'project originators' and thus take on 'client duties'.
- No legal duty under CDM regulations to review assessments, monitor performance of appointees or continued adequacy of the health and safety plan during the construction phase.
- ACOP provides examples of relevant information to be provided to the planning supervisor early enough for implications to be assessed by the designer.

Health and safety plan and file

Inclusion of unnecessary and irrelevant information that can detrimentally affect the ease of identification and significance of crucial details.

- Appendix 3 contains a list of matters that should be included or addressed within the health and safety plan, where relevant to the work proposed. This is in tabular format with those of the pre-tender stage alongside those of the construction phase under the same topic headings, to illustrate and compare the appropriate requirements for each.
- Appendix 4 has a similar list of contents of the health and safety file and the ACOP also lists those topics that do not need to be included.

Consultation of construction workers

Decisions on health and safety arrangements are often made without consulting workers.

ACOP extols the necessity for and the benefits of obtaining feedback from workers' formal or informal safety committees or representatives, by addressing common problems, reviewing accidents and near misses, and identifying and considering how risks should be addressed on site.

Town and country planning in England

- Development applications and fees
- Appeals and called-in applications
- Development plans and monitoring
- General Development Order and Use Classes Order
- Planning policy guidance notes and circulars
- Listed buildings and conservation areas
- Environmental impact assessments

Legislation and regulation

Legislation and regulation

Development applications and fees

Building, engineering, mining or other operations in, on, over or under land, including demolition, all require development permission, but many minor matters have permitted development rights (General Permitted Development Order) and demolition consent is only required in specific cases (Circular 10/95).

Material changes of use of land or buildings require permission but many have deemed permission (Use Classes Order and General Permitted Development Order).

Buildings which are listed as of special historic or architectural interest (Grades 1, 2* and 2) may not be altered (or demolished) in any way that might affect their character as a listed building without listed building consent. Unlisted buildings in a conservation area may not be wholly demolished without Conservation Area Consent. Scheduled monuments may not be altered at all without consent. Advertisements may not be erected without Advertisement Control Consent but many have deemed consent (Advertisement Control Regulations). Applications should be dealt with in eight weeks by the Local Planning Authority or in 16 weeks if an environmental statement is submitted.

A fee is payable for every planning and advertisement application but the Application Fee Regulations allow for some exemptions or reductions normally related to the type of applicant, or where an earlier application has been submitted and refused or withdrawn. There are no fees for entering planning appeals but there are for enforcement appeals.

Appeals and called-in applications

All types of application give rise to a right of appeal against refusal, non-determination or conditional approval. Appeals are dealt with by the Planning Inspectorate, an executive agency of central government reporting to the Office of the Deputy Prime Minister (ODPM). Appeals must be made within a certain time from the date of the decision of the relevant council (or within six months of when they should have made a decision), and can only be made on specific appeal forms available from the Planning Inspectorate.

There are three types of appeal procedure – public inquiry, hearing or written representations – secondary legislation, in force since 1 August 2000, governs the timetabling leading to the inquiry, hearing or site inspection, but there is no time limit on the decision-making process itself. DETR Circular 5/2000 summarises the appeal procedures.

Called-in applications are rare but have the effect of opening up the application to public debate at an inquiry, with the decision being made by the ODPM and not the Local Planning Authority.

Development plans and monitoring

Where the development plan is material to the development proposal and must therefore be taken into account, Section 54A of the 1990 Planning Act requires the application or appeal to be determined in accordance with the plan, unless material considerations indicate otherwise. In effect, this introduces a presumption in favour of development proposals, which are in accordance with the development plan (Planning Policy Guidance Note No.1: General Policy and Principles).

The primacy of development plans is now enshrined in the planning system and ignoring the opportunity to make representations on unitary,

Legislation and regulation

local or structure plans can seriously damage development prospects and limit the opportunities available for redevelopment or changes of use. Similarly applicants, or their agents, must be aware of the relevant policies in the development plan before entering any application. The monitoring of development plans is therefore an essential element of both sound estate management and of preparation for development.

General Development Order and Use Classes Order

The Town and Country Planning (General Permitted Development) Order 1995 (amended 2001)

The GPDO gives a general permission for certain defined classes of development or use of land, mainly of a minor character. The most commonly used class permits a wide range of small extensions or alterations to dwelling houses, and there are many categories that allow for development that would otherwise require planning permission. There are allowances for developments which would also require listed building, conservation area or advertisement control consent, but this does not remove the need for such consents.

The general permission that the GPDO grants for a particular development or class of development may be withdrawn in a defined area by an Article 4 Direction made by the local authority or the ODPM.

The Town and Country Planning (General Development Procedure) Order 1995

This order came into force in June 1995, as did the GPDO, and details procedural matters on applications, including a full list of statutory consultees, as well as requirements for registers of applications, specimen copies of forms and guidance on appeal submissions.

The Town and Country Planning (Use Classes) Order 1987 (as amended)

Section 55 of the 1990 Planning Act provides that changing certain uses does not constitute development under the Act and this includes any change of use of land or buildings from one use to another within the same class of the Use Classes Order. Other changes of use are not necessarily development, only being so when the change is material; normally when changing from one use class to another outside the allowances in the UCO or GPDO, or when the new use is of a different character to the old use.

A 1993 change to the UCO removed hostels from the hotels and hostels use class (Class C1) and there is now much case law clarifying changes of use. Circular 11/95 of the DoE makes it clear that there is a presumption against imposing planning conditions designed to restrict further changes of use which, by virtue of the UCO, would not otherwise constitute development.

Since the UCO was published in 1987, many organisations have produced comparative tables of permitted changes but in view of the refinement and updating of the UCO these are now suspect and each case should be considered in the light of the current legislation. The courts have held that it is not necessary to go to extreme lengths to identify a class for every use and the UCO lists a number which are not to be regarded as being in any class of the order. This does not necessarily mean that a change to, or from, such a use, to a use within a specific class will always be a material change of use requiring planning permission.

Legislation and regulation

Uses which have an uninterrupted history of ten years, but which do not have planning permission, may now be the subject of an application for certification of the lawful use commensurate with a planning approval. Established Use Certificate legislation requiring a pre-1964 commencement of the use has been withdrawn.

Planning policy guidance notes and circulars

In January 1988 the Department of the Environment introduced a new set of guidance in the form of planning policy guidance notes (PPGs). As at November 2002 there were 25 of these. The PPGs are to be taken into account by local authorities as they prepare their development plans, and they may be material to decisions on individual planning applications and appeals. Guidance in the PPGs can be considered as particularly material if the development plan does not reflect the policy direction in them.

PPG Number	Title	Date
1	General policy and principles	1997
2	Green belts	1995
3	Housing	2000
4	Industrial/commercial development and small firms	1992
6	Town centres and retail developments	1996*
7	The countryside – environmental quality and economic and social development	1997*
8	Telecommunications	2001
9	Nature conservation	1994
10	Planning and waste management	1999
11	Regional planning	2000
12	Development plans	2000
13	Transport	2001
14	Development on unstable land	1990
14a	Annex 1: landslides and planning	1996
15	Planning and the historic environment (see circular 1/2001)	1994*
16	Archaeology and planning	1990
17	Sport and recreation	2002
18	Enforcing planning control	1991
19	Outdoor advertising control	1992
20	Coastal planning	1992
21	Tourism	1992
22	Renewable energy	1993
23	Planning and pollution control	2002
24	Planning and noise	1994
25	Development and flood risk	2001

* = amended by ad hoc statements

(r) = under revision

Since May 1996 The Welsh Office has prepared technical advice notes and planning policy guidance just for Wales. The Scottish Office publishes planning advice notes. Regional planning guidance is now contained within RPGs and mineral planning guidance in MPGs for England.

Legislation and regulation

Circulars

Circulars are now used primarily to publicise technical and administrative changes but there are many that are still particularly relevant, notably:

- Circular 13/87: The Use Classes Order
- Circular 06/98: Affordable Housing
- Circular 5/94: Planning out crime
- Circular 9/95: GDO Consolidation
- Circular 10/95: Demolition
- Circular 11/95: Planning Conditions
- DETR Circular 2/99: Environmental Impact Assessment
- DETR Circular 5/2000: Planning appeals
- DETR Circular 1/2001: Heritage Applications
- GOL Circular 1/2000: Strategic Planning in London.

The ODPM has introduced a housing density circular concerning the south-east (1/2002).

Greater London

The Mayor of London is responsible for certain planning functions and has published draft planning guidance and strategies.

Listed buildings and conservation areas

'Planning and the historic environment' (PPG15) replaced policy guidance in Circular 8/87. The PPG places greater stress on keeping historic buildings in active use and urges planning authorities to be flexible and imaginative in their approach, to achieve the right balance between protecting a building's special character and adapting it for new uses.

Listed buildings in England

The principal current legislation is the Planning (Listed Buildings and Conservation Areas) Act 1990. This sets out the statutory framework relating to listed buildings and conservation areas.

Listed Building or Conservation Area Consent is not required for a material change of use. Change of use is covered under the Planning Act 1990.

Grade 1:
Buildings of exceptional interest (about 2% of listed buildings)

Grade 2*:
Particularly important buildings of more than specialist interest (about 4% of listed buildings)

Grade 2:
Buildings of special interest which warrant every effort being made to preserve them

Listing

The basic criteria for listing are set out in PPG15 and while 'spot listing' will still be considered for individual buildings overlooked or under threat, by reference of details by any member of the public to the ODPM/English Heritage, future research to identify potential buildings for listing will alter.

Legislation and regulation

The list reviews will concentrate on finding the best examples of types and periods of buildings under-represented in the current list.

What is a listed building?

Any structure or erection, and any part of a building including any object or structure within the curtilage that forms part of the land and did so before 1 July 1948.

Material change of use

It is accepted that new uses for old buildings may often be the key to their preservation. In some instances, it may be appropriate that control should be relaxed where this would enable historic buildings to be given a new lease of life.

The best use for an historic building is obviously the use for which it was designed and, wherever possible, the original use should continue.

Works for which listed building consent is necessary

Demolition of a listed building, or its alteration or extension in any manner that would affect its character, requires consent. It should be noted that the setting of an historic building is considered of great importance and an essential feature of its character. Consent is not required for works of repair that are on the basis of like-for-like materials and exact details. However, this is a grey area.

Application for listed building consent

Applications for listed building consent should be made to the local planning authority.

Listed building consent decisions

The local planning authority has eight weeks in which to consider an application. This can be extended by agreement with the applicant or revision of proposals, etc.

The legislation requires all listed building consents to have conditions, even if only time/recording conditions.

Listed building consent for works already executed

Listed building consent may be sought even though the works have already been completed. If consent is granted this is not retrospective; the works are authorised only from the date of consent.

Conservation areas

Local authorities have a duty imposed by section 71 of the Listed Buildings Act 1990 that they must regularly formulate and publish proposals for the preservation and enhancement of conservation areas, after consultation with local people at a public meeting. The advice insists that simply designating a conservation area will not ensure its protection and that planning authorities should analyse what makes an area special and develop policies to protect it, in consultation with local residents and businesses.

In 1997, the House of Lords ruling in the Shimizu case changed the previously understood meaning of 'listed building'. This had implications for both listed building and conservation area controls. Before Shimizu, the demolition of a listed building was taken to cover either its total or partial demolition. The ruling enables partial demolition of unlisted buildings in

Legislation and regulation

conservation areas, but the government is proposing legislative charges with potential to reverse the decision.

Other powers

The guidance gives powers to councils to control some minor developments in conservation areas including alterations to roofs, doors and windows, so as to prevent damage to the area's special character and to avoid councils having to pursue individual Article 4 directions. From October 1994 many churches, previously exempt, were brought within conservation controls although christian churches in active use still have ecclesiastical exemption.

Following publicity for the English Heritage list of battlefields, PPG15 reminds developers and councils to also protect wider features of the historic environment such as important gardens and parks, which are now separately listed and graded. World Heritage Sites have a high measure of protection.

Preservation presumption

PPG 15 re-affirms the presumption in favour of preserving listed buildings of special architectural or historic interest and a presumption in favour of retaining unlisted buildings which make a positive contribution to the character or appearance of a conservation area.

Devolution of power

English Heritage has withdrawn from the day-to-day control of Grade 2 listed buildings in the City of London, Westminster and several of the London boroughs. In the remainder of England, English Heritage are consulted primarily on Grade 2* and Grade 1 buildings.

Major alterations to any listed building must be referred to English Heritage (s15 of the Listed Buildings Act).

Environmental impact assessments

The principal environmental impact assessment (EIA) regulations are the Town and Country Planning (Environmental Impact Assessment) (England and Wales) Regulations 1999, as amended in 2000. Where a particular proposal is likely to have a significant environmental effect then a local planning authority may request an EIA for developments of the type listed in Schedule 2. Schedule 1 developments always give rise to EIAs. An applicant may apply for a screening direction (through a speedy process) to the ODPM if the applicant believes the EIA is unnecessary. A 16-week decision period applies to any EIA application, running from the deposit of the EIA with the local authority. From June 1995 any development requiring EIA loses the benefit of permitted development. See DTLR Circular 2/99.

Legal and lease

Legislation and regulation

- Dilapidations
- Latent Damage Act 1986
- Discovery of building defects – statutory time limits
- Expert witness
- Dispute resolution
- Adjudication under the Scheme for Construction Contracts - how to get started

Legislation and regulation

Dilapidations

The principle

Dilapidations is part of a legal procedure and the fundamental purpose of a Schedule of Dilapidations is to identify any breaches of covenant within a lease. The allegation of a breach of contract is the first step in the legal process. Consequently, for a Schedule of Dilapidations to have worth and perform the function for which it is drafted, it is necessary that the schedule is enforceable in a court of law.

The legal remedy for breach of contract is often a claim for damages and therefore the Schedule of Dilapidations is usually prepared as a claim for the cost of works.

Reform

New Civil Procedure Rules came into force in April 1999. A protocol has been drawn up by the Property Litigation Association (the most recent draft is dated July 2002) and it is hoped it will be submitted to the Lord Chancellor's Department to be approved under the Civil Procedure Rules (CPR). The underlying intention of CPR is to increase the number of pre-action settlements, reduce court time, expense and ultimately the extent of litigation. The Civil Procedure Rules have made a number of changes.

- ❖ Pre-action offer - this can be made by either party pursuant to Part 36 of The Civil Procedure Rules.
- ❖ All claims must be accompanied by a statement of truth signed by the claimant.
- ❖ Expert witness - the role has fundamentally changed under CPR. The expert's function is to give an independent expert view; he owes a duty to one body only, and that is the Tribunal before whom he appears. He is not there to create an argument in support of the party that appointed him.

The protocol

Although the protocol has not yet been approved under CPR, the courts may treat it as the normal and reasonable approach to pre-action conduct, and non-compliance might bring sanctions against the party concerned.

The protocol suggests a time limit of two months, after the end of the term for the landlord to have served the Schedule of Dilapidations. The tenant then has two months to respond and thereafter the respective surveyors should meet on site to discuss the claim, and seek to agree as many of the items in dispute as possible.

An early assessment of the claim is essential.

The claim should be drawn up as a separate document and should identify how it has been compiled, give a summary of the facts that the claim is based on, the VAT status of the landlord, all supporting documents and the date by which the tenant should respond.

The protocol clause 4.1.2 states that:-

- ❖ If the landlord has carried out the work then no Section 18 (1) Valuation is required.
- ❖ If the landlord intends to carry out the work, then it must state when the works are due, what steps have been taken towards doing them, and in most cases, it must provide a Section 18 (1) Valuation.
- ❖ If the landlord does not intend to carry out the works then it should provide a Section 18 (1) Valuation.

Legislation and regulation

Stages in the dilapidations process
Stage 1 – preparation

Obtain and appraise all relevant documentation including:
- leases;
- licences to alter;
- schedule of condition;
- side letters;
- photographs;
- fit out specifications;
- agent's letting brochures;
- any statutory notices served;
- deeds of variation;
- schedules of landlord and tenant's fixtures and fittings;
- details of outstanding service charges; and
- any rent deposit agreements.

This list is by no means exhaustive.

Stage 2 – the inspection

This must be comprehensive and thorough and include specialist professions if deemed necessary, such as a structural engineer or mechanical or electrical consultant.

Establish the original condition at the beginning of the term, and standard of repair that the tenant is required to undertake and identify the remedial work. Take into account the age, character and locality of the premises when let (Proudfoot Vs Hart 1890).

Include all measurements to aid calculation of the remedial works and use as proof as required at a later date.

Stage 3 – preparation of the Schedule of Dilapidations and claim

The schedule should contain the information shown in the first five columns below. The additional columns relating to the tenant's and landlord's comments turn the Schedule of Dilapidations into a Scott Schedule used as a negotiating tool between respective surveyors.

ITEM	CLAUSE No.	BREACH	REMEDIAL WORK REQUIRED	LANDLORD COST (£)	TENANT'S COMMENTS ON		LANDLORD'S COMMENTS ON	
					BREACH & REMEDY	COST (£)	BREACH & REMEDY	COST (£)

The Schedule of Dilapidations should be accompanied by a claim letter, which must include:
- landlord's and tenant's name and address;
- a clear summary of the facts on which the claim is based;
- the Schedule of Dilapidations (a separate document);
- any documents such as invoices and evidence of costs;

- ❖ confirmation that the landlord and advisors will attend meetings;
- ❖ a date by which the tenant should respond; and
- ❖ a summary of the claim including:
  - cost of works
  - preliminaries
  - overheads and loss of profit
  - surveyors fees for preparing the Schedule (quantified and substantiated)
  - loss of rent
  - loss of service charge
  - surveyors fees for negotiating a settlement (projected)
  - if a Section 18 (1) cap applies then this should be shown.

Stage 4 – the response and negotiations

The Schedule of Dilapidations should be served within a reasonable time before the termination of the tenancy but not more than two months afterwards.

If a notice from the landlord has to be given to the tenant for reinstatement items then this must be served within a reasonable period before the end of the tenancy so that the tenant is still able to carry out these works in time.

Electronic copies of the Schedule of Dilapidations should be provided to facilitate easier negotiation, preferably in a Scott Schedule format as shown in the diagram on the previous page.

Following submission of the claim, the tenant must respond within a reasonable period, usually two months.

Surveyors should meet within one month of service of the Schedule of Dilapidations on a without prejudice basis preferably on site to establish the facts. If further meetings are necessary a strict timetable should be adopted.

Experts of respective parties in their like disciplines should also meet within one month of the tenant's response.

Proceedings should not be issued less than one month after the meeting of experts, or three months after serving of the schedule (whichever is the earlier).

Valuation – diminution in value

Where a building is in disrepair at the end of the term, Section 18 (1) of the Landlord and Tenant Act 1927, limits the landlord's claim for damages for breach of a repairing covenant. Other breaches are covered by Common Law Principles usually related to a landlord's loss as referenced by diminution in value.

There are two parts to Section 18 (1):-

- ❖ The first limits the claim to the amount that the value of the landlord's reversion is diminished by breaches of the covenant to repair. The landlord cannot recover more than it has cost, in terms of the loss caused to the value of the property. This is the diminution in the property's reversionary value, caused by the disrepair.
- ❖ The second part states that no damages are recoverable, if it can be shown that on expiration of the lease the premises would be demolished or altered to the extent that would render valueless the repairs in question.

In order to calculate the diminution, two valuations are required. First to compare the property's value in poor repair and second to look at the value

Legislation and regulation

the property would have achieved in the open market if it had been maintained by the tenant.

The second valuation is a hypothetical assessment, as the premises would not currently be in repair. It must consider all potential uses for the premises, as it may be more valuable as a redevelopment site or for an alternative purpose. It must also take into account the extent that an alternative or change of use would affect individual elements of the schedule. For instance repairs to the external areas may still need to be carried out.

Redevelopment may well render valueless any repairs that may be required by the lease covenants, and subsequently the landlord's intentions for the property must be considered at the expiration date. In the case of Firle Investments Limited v Datapoint International Limited (2000), the landlord considered refurbishment of an office building following the expiry of the 25 year lease. The judge considered whether at the expiration date, the landlord's refurbishment intentions had moved "out of the zone of contemplation - out of the sphere of the tentative, the provisional and the exploratory - into the valley of decision," as described by Asquith L J, in Cunliffe v Goodman, 1950.

The protocol clause 4.1.2 (described under 'The protocol' above) requires a Section 18 (1) Valuation to be carried out in some situations. It also states that the landlord's claim must include a summary of the landlord's intentions. Post-termination events are admissible as a means of assessing the diminution in value of the reversion at the end of the term. If the courts are unable to act on fact, then the requirements of a hypothetical purchaser and transaction will be assessed.

Dilapidations v The Disability Discrimination Act (DDA) 1995

By October 2004, a provider of goods and services to the public will be obliged to take steps to change physical features within a premises for disabled people.

This will include removing, altering, or providing alternatives that would enable disabled people to make use of the service. The responsibility and duties in the DDA can rest on either a landlord or a tenant.

If a tenant, either in a single let building, or a multi-let building, provides a service to the public, the duties fall on the tenant.

If a landlord provides a service to the public, then they are responsible for any common parts of the building, whether it is a multi-occupied office building, or a shopping centre. However, it is not clear who is responsible for the common parts in the case where the public only visit the building at the invitation of the tenant.

Where the responsibility for compliance lies with the tenant, the tenant is not under a statutory requirement to the landlord for making adjustments to the premises.

It is unlikely that the landlord would be able to require a tenant to undertake works to comply with its duties under the DDA, during, or as part of a dilapidations requirement at the end of the lease.

Service charge

One of the most important aspects of maintaining a landlord's investment in a property, is to ensure that it is kept in repair. In a multi-occupied building, let on full repairing terms, the landlord will seek to recover costs of all repair and maintenance work from various tenants through a service charge.

An important recent case, Fluor Daniel Properties Limited v Shortland Investments Limited (2001), encompasses some already established

Legislation and regulation

principles relating to service charge recovery.

In order for a landlord to carry out repairs to a multi-occupied property, and recoup the cost from the tenants, he must establish that the item in question is out of repair, and to do this, a number of circumstances must be taken into account. These include the nature and life span of the building, the length of lease, and the extent of the defect. The courts also state that a landlord must have regard to the benefit that a tenant would derive from the repairs.

The question of standard relies on a number of circumstances as laid down by Proudfoot v Hart (1890). The standard must have regard to the age, character and locality of the building and whether the condition of the subject matter is reasonably acceptable to a reasonably minded tenant, of the kind likely to take on a lease of the building or part thereof.

The landlord must have proper regard to the interests of the tenants, and in some circumstances may carry out improvements. As the environment in which the building operates is improved, a reasonable tenant may expect features to be improved such as high speed lifts, or better performing air conditioning.

The landlord must act reasonably and the obligation to repair must not be looked at on its own. In every instance it is a question of degree.

Break clauses

These are covenants within a lease that give the option for the landlord or the tenant to bring the lease to an end before the contractual expiry date. It is important to establish whether they are condition precedent or not.

Where they are conditional, the tenant may be required to comply with certain conditions before the break option can be successfully exercised. This may include full compliance with all repairing obligations within the lease. Failure to comply may result in the lease continuing for the remainder of the term.

VAT

Protocol, Clause 4.2, requires that the VAT status of the landlord should be stated. VAT in respect of dilapidations is a complex subject with its own set of rules, regulations and case law. Opinions are usually divided and much will depend on the circumstances.

Where a tenant is making a payment to a landlord in full and final settlement of its dilapidation liabilities, under Customs and Excise rules, the payment is not a 'taxable supply' for the purposes of VAT. This is because Customs and Excise deem the payment to be one of damages and not a supply of something. VAT may be payable if the landlord passes the payment on to an incoming VAT registered tenant.

Complying with European Community Law, the Finance Act 1989 introduced major changes that included giving UK payers of VAT the option to pay VAT on supplies relating to an interest in commercial land. Where a person or company is registered for VAT there is a special statutory exemption to the charging of VAT on such supplies. The tax payer can waive this exemption if they desire. To do this they must notify HM Customs and Excise. For clarity, if a landlord has elected to waive exemption on a building, he must charge VAT on such items as rent received (ie supplies that the tenant is receiving). In this instance, it will not be appropriate to include VAT as part of a dilapidations claim. If the VAT exemption has not been waived then VAT need not be charged by the landlord on supplies, and VAT may form part of a dilapidations claim.

So the first issue to establish is whether the landlord is registered for VAT. Is the landlord required by VAT law to charge VAT on its normal business transactions, such as rent receipts. If so, has it elected to waive VAT exemption on the specific building that is subject to the dilapidations claim?

Legislation and regulation

Once this has been established, a simplified VAT analysis might follow the steps shown in the diagram.

The protocol requires that the claim establishes what the landlord's intentions are, and states the landlord's and/or the demised property's VAT status. A basic understanding of the VAT issues is therefore necessary, if the requirements under the protocol for an accurate initial claim are prepared.

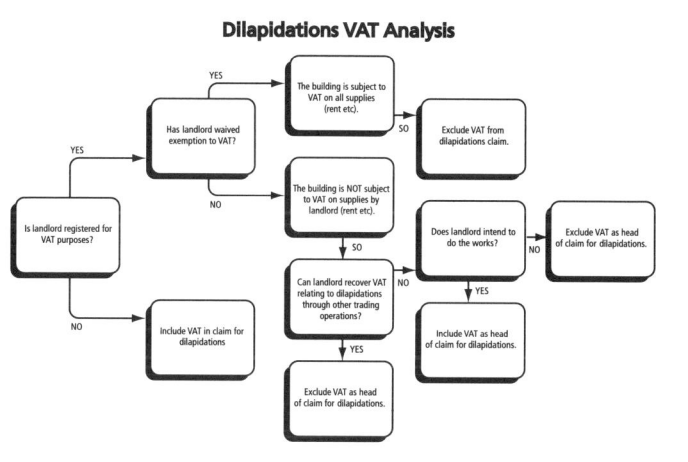

Dilapidations VAT Analysis

```
Is landlord registered for VAT purposes?
  YES → Has landlord waived exemption to VAT?
          YES → The building is subject to VAT on all supplies (rent etc). SO → Exclude VAT from dilapidations claim.
          NO → The building is NOT subject to VAT on supplies by landlord (rent etc). SO → Can landlord recover VAT relating to dilapidations through other trading operations?
                YES → Exclude VAT as head of claim for dilapidations.
                NO → Does landlord intend to do the works?
                       YES → Include VAT as head of claim for dilapidations.
                       NO → Exclude VAT as head of claim for dilapidations.
  NO → Include VAT in claim for dilapidations
```

Summary of important statutes

Landlord and Tenant Act 1927 Section 18 (1)
Limits the cost of a claim for breach of covenant.

Law of Property Act 1925 Section 146
Prescribes the form of notice for re-entry for forfeiture.

Law of Property Act 1925 Section 147
Provides relief for tenants on long leases in respect of internal redecorations.

Leasehold Property (Repairs) Act 1938
Gives protection to certain tenants in respect of Section 146 Notice (ii) above.

Defective Premises Act 1972
Provides that a landlord shall be liable for lack of repair in cases where he/she knew or ought to have known of the defect.

The Civil Procedure Rules 1998
Introduced by Lord Woolf, they provide rules and practice directions for dispute procedures.

Legislation and regulation

Summary of important case law

There have been a number of leading decisions relating to dilapidation law in the last few years. The following cases give an indication of developing areas of law.

Scottish and Mutual Assurance Society Limited v British Telecommunications plc (1999) (E.G.C.S 43)

Section 18 (1) of the Landlord and Tenant Act 1927 Part II.

Loss of rent.

Notice for reinstatement of alterations.

Shortlands Investments Limited v Cargill plc (1995) (E.G.L.R 51)

Section 18 (1) of the Landlord and Tenant Act 1927 Part II.

Trane (UK) Limited v Provident Mutual Life Assurance Co Limited (1995) (W.G.L.R78)

Compliance with conditions of break clauses.

Jervis v Harris (1996) (E.G.L.R 78)

Use of provision for landlord's re-entry.

Extent of recovery of expenditure.

Mannai Investments Co. Limited v Eagle Star Life Assurance Co Limited (1997) (E.G.L.R 69)

Accuracy of notice showing intent to break tenancy.

Credit Suisse v Beegas Nominees Limited (1994) (4All,ER803)

Establishing the interpretation of the repairing covenants and the different types of obligation placed on the tenant.

Latent Damage Act 1986

The Latent Damage Act 1986 enhances case law relating to certain cases of negligence, by imposing statutory limits in relation to the time in which cases may be brought to court if a defect is found.

The Act applies specifically to those cases of negligence that relate to latent damage and not to personal injury. In addition, recent case law suggests that in the meaning of the Act, negligence covers only a breach of a tortious duty of care and not a contractual duty of care.

Previous statute exists indicating time limits in which legal action must commence; section 14 of the Limitation Act 1980. The 1986 Act enhances section 14 by limiting any claims to six years after the damage occurred. This, however, can be extended under section 14A by a further three years from the date when the defect is discovered. In all cases however, a 15-year time limit is placed under section 14B of the Act. When considering claims for latent defects, the most important points to establish are the dates from when the negligence occurred and the cause of the negligence. For example, the negligence may have been caused by defective construction, design, or a combination of both. The date on which the defect technically arises may be confused by the ongoing contractural state of the project, ie whether a final certificate has been issued or a defect liability period is still active. In certain cases proceedings may be delayed on the basis that the damage has not yet occurred, but is in fact imminent.

Legislation and regulation

Discovery of building defects – statutory time limits

The Limitation Act 1980 ('the 1980 Act') as modified by the Latent Damage Act 1986 (see above) lays down the periods within which proceedings to enforce a right must be brought. Upon the discovery of a defect in a building or structure, possible claims against the designers and constructors of that building or structure could arise in contract, in tort or under statute.

Contract

To bring an action for breach of a simple contract, court proceedings must be commenced within six years of the date on which the breach of contract occurred. If the contract has been completed under seal, rather than under hand, then the time limit prescribed by the 1980 Act is 12 years. Under the Companies Act 1985 (as amended) the 12-year limitation period also applies to companies when a contract is signed as a deed by two directors or a director and a company secretary.

Tort

The general time limit for actions in tort is six years from the date when the damage was suffered, with the exception that a time limit of three years applies to personal injury actions involving negligence. This period runs from the date of the accident concerned, although special rules apply for illnesses, which may not manifest themselves for many years after exposure to their cause, for example, asbestos.

Also, for negligence claims involving latent damage, the time limit laid down by the Latent Damage Act 1986 for commencing proceedings is six years from the date the damage was suffered. However this period can be extended for a further three years from the 'starting date'. The starting date is defined by the Latent Damage Act 1986 as the date on which the plaintiff first had both the required knowledge and the right to bring an action. The Latent Damage Act also includes a 'long-stop' provision preventing the instigation of any proceedings after the expiry of 15 years from the date of the negligence concerned.

Statute

Where a right derives from a breach of statutory duty, reference should, in the first instance, be made to the particular statute concerned, which may specify a time limit for the commencement of proceedings. The Defective Premises Act 1972 provides a particularly germane example. If the statute concerned is silent as to the time limit, then a period of six years will generally apply. If an action is brought in respect of a defective product, under the Consumer Protection Act 1987 there is a cut-off point of ten years from the 'relevant time' (usually the date of a supply).

It should be noted that the time limits for both tortious and contractual claims might be postponed where there is concealment, mistake or fraud.

Expert witness

The Civil Procedure Rules (CPR) which came into force on 26 April 1999 have changed the way the courts go about their business by allowing them to manage the cases before them. Part 35 of the new CPR deals with expert witnesses. The rules do not create a new breed of expert witness, but permit the courts to take more control over experts and their evidence.

Legislation and regulation

It has always been the case that experts are independent/impartial and their duty is to assist the court in reaching its judgement. The new CPR states, "it is the duty of an expert to help the courts on matters within his/her expertise. This duty overrides any obligation to the person from whom he/she has received instructions or by whom he/she is paid". The expert should therefore:

- be independent/impartial;
- state any reservations about the case he/she is instructed upon;
- identify areas outside his/her expertise;
- consider all material facts in his/her report and state all facts and assumptions, upon which his/her opinion is based;
- state where necessary that it is only a provisional report, because only limited information/data was available when it was compiled;
- advise instructing solicitors if further information, ie the other expert's report, changes his/hers/your opinion;
- make available all documents referred to in the report, ie survey reports, plans, calculations, photographs etc; and
- keep the report as brief as possible, but without losing the reasoning and conclusions, upon which his/her opinion is based.

Taking instructions

It is essential that the expert witness is briefed/instructed correctly and in detail as to what his/her expert evidence will cover. The brief/instructions will need to be disclosed in the expert's report. Instructions should come from the solicitor acting for the party for whom you are going to give expert evidence. The brief/instructions may need to be amended as the case develops.

In large, complex cases there may be a number of experts. The new CPR also allows the courts to order that each party just has one lead expert, who may be the only one called to give evidence, but may incorporate in his/her report the work of other experts under his/her control. The brief/instructions should cover the following:-

- The nature and subject matter of the proceedings, the amounts involved and the main issues identified.
- Details of all other parties their advisors and experts.
- What precisely is the expert witness being asked to do.
- Details of all members of the team including any other experts.
- How the role fits in with the roles of the other professional experts and the case generally.

When taking instructions it is also essential to establish the timetable and critical dates. The courts are increasingly hard on parties who fail to comply with the timetable, which can result in the expert witness incurring costs and penalties and the possibility of his/her evidence being excluded.

At the time of briefing/instructions, then the matter of fees and payment needs to be addressed to determine on what basis expert witness' charges will be made, ie fixed fee, hourly rate or otherwise. What are the payment terms? How will disbursements be dealt with? Does the client require the expert witness to keep detailed records of time spent? Are cancellation fees necessary? If costs were assessed, how will this affect fees, ie will the fee be restricted to the amount certified by the court?

Legislation and regulation

The written report

There are a number of model forms of report, for example, the model form produced by the Academy of Experts. The CPR lays down formal requirements in the practice direction that accompanies part 35 of the new rules. It is extremely useful to discuss the format of the report and its contents with the instructing solicitor.

Under the new rules, there will need to be a statement setting out the brief and instructions given.

The report is to be written in the first person and it is an individual who prepares the report and not the company or firm.

Under the CPR the report should be addressed to the court, but you should seek instructions from the solicitor.

Although the report should be as brief as possible, accuracy should not be sacrificed to brevity.

The expert witness must be able to substantiate each and every sentence of the report and highlight any areas where his/her opinions are based on inadequate factual information. It is not the expert's role to make or advance legal arguments.

The report should contain his/her curriculum vitae (CV).

He/she should seek guidance from the solicitors on including hearsay evidence in the report. It is compulsory under the new rules to include a declaration in the report that the expert understands that his/her duty is to the court and that he/she has complied with that duty. A statement of truth must also verify the report. The wording to be included within the expert's report immediately before the signature is as follows:

> I believe the facts as stated in this report are true and that the opinions I have expressed are correct.

Under the CPR, each party has 28 days after receipt of the opposing expert's report to put written questions. Unless the court gives permission for more general questions, these can only be for the purpose of clarifying the report.

Without prejudice meetings

Without prejudice meetings between experts are critical and can win or lose the case. Hold as many as you need. An expert has a duty to cooperate with the expert of the other party.

Meetings between experts can produce information that is extremely valuable when preparing for trial. As the meeting is 'without prejudice' it may provide an opportunity to ascertain what the other side considers to be the strengths and weaknesses of its case.

The court, under the CPR, may require the experts to produce a joint statement from 'without prejudice' meetings.

Giving evidence in court

If the case gets to a hearing, then the expert witness will be required to give evidence. The stages of examination of the evidence will be:

- examination in chief;
- cross-examination;
- re-examination; and
- questions from the judge.

A few helpful hints in giving evidence are listed below:-

- Take time, don't rush.
- Succinctly answer only the questions that are asked.
- Use plain language.
- Do not digress from the question asked.
- Do not act as advocate.

Legislation and regulation

- If the question is not understood nor heard – say so.
- Know the report 'inside out'.

Cross-examination will challenge credibility; so consider the following:-

- Do not feel obliged to fill a silence.
- Do not be afraid to answer the same question again and again.
- If asked a closed question, then state that there may not be a yes or no answer.
- Do not be rattled by a number of quick fire questions.
- Do not argue with counsel.
- If he/she becomes aggressive, stay cool.
- Beware of the important question slipped in among a number of trivial questions.

Giving evidence is probably one of the most stressful experiences of one's professional life. But be surprised at just how rewarding it can be.

Footnote

The above notes only deal briefly with the new Civil Procedure Rules. These notes should not be considered as comprehensive text. The role of the expert is evolving through the interpretation of the CPR in case law and any expert should ensure that he/she fully understands that role in the light of the current law.

Dispute resolution

The property and construction industry have for many years indulged in disputes as a normal part of complying with the duties and liabilities that they owe to each other. This has given rise to significant concern both inside the industry and with its customers, our clients.

Cost and time factors and the often unsatisfactory outcome of the dispute process has gained the industry a reputation for inefficiency and gives it a poor image in the minds of the wider commercial community.

While it may be the case that the subject matter of our industry is inherently disposed to argument and disagreement, we do owe our clients an alternative way of resolving matters that allows them to achieve a quick, certain, fair, understandable and cheap dispute resolution process. Certainly, some disputes are complex and involve substantial sums and these will require due process if the parties' proper interests are to be protected. Other disputes will require different procedures. A wide range of procedures is now available to the industry, which will allow disputants to match the dispute with the resolution process.

More important, there is a climate created by statute, the judiciary and users who have expressed a determination to 'do things differently'.

Recent years have seen the publication of the Latham Report, which made many valuable comments and suggestions on how the industry may cut down this flow of disputes by adopting more collaborative working practices. The inevitability of disputes was recognised by the major advances in dispute resolution afforded by the Construction Act 1996. After a slow start, it is now playing a major role in adjudicating disputes. Undoubtedly, the support of the judiciary in upholding adjudicators' decisions will continue to play a strong part in the success of this procedure.

The introduction of the Civil Procedure Rules (CPR) have made far ranging changes to the civil justice procedure by simplifying and demystifying the process. Important initiatives were made in the use, often on an accelerated basis, of alternative dispute resolution procedures as a means of settlement before trial.

Legislation and regulation

Disputants and their advisors now have a wide variety of dispute resolution mechanisms that they can select to resolve their disputes. The procedures are either relatively new, like adjudication, or have been overhauled and made more commercially attractive by recent legislation and other reforms.

The principal dispute resolution procedures are:

Adjudication

This process is now enshrined in the Construction Act, which refers a wide variety of the disputes arising under construction contracts to adjudication. The adjudication procedure laid down must comply with all the basic minimum criteria set down in the Construction Act, and if it does not then the government has imposed the Scheme for Construction Contracts on disputants to govern the adjudication. The Construction Act does not deal with residential disputes, but the JCT Consumer Contracts contain adjudication clauses. Adjudication is a popular method of dispute resolution not least because an adjudication award has to be made within 28 days of the case being referred to an adjudicator.

Arbitration

Arbitration is based on the contractual relationships between the parties themselves and between them and the arbitrator. There are resemblances to litigation, but arbitration should be increasingly flexible and cost effective. The increasing establishment of the new procedures under the 1996 Arbitration Act has given the arbitrator wide powers to resolve disputes without unnecessary cost or delay, and in a fair manner without undue interference from the courts.

Early Neutral Evaluation (ENN)

Technology and construction court judges are prepared to arrange a short hearing of a case, or specific issues in it, on a without prejudice basis and give preliminary views on its merits, as an aid to settlement discussions between the parties. If a judge determines a particular issue by ENN the parties are free to agree whether or not they will be bound by it. If the ENN does not result in settlement, the case can proceed to trial but will be heard by another judge with no knowledge of the outcome of the ENN.

Independent expert

This forms a very valuable means for the speedy resolution of technical disputes. The procedure is generally straightforward and flexible. Issues in dispute are referred to an expert to decide using his/her or her own professional expertise or judgement. It has been used for many years quite successfully in rent review matters, but has much wider application to technical disputes. If the parties agree to be bound by the expert's decision it cannot be appealed unless there is misconduct on the part of the expert.

Litigation

The CPR has led to more efficient running of cases both in terms of costs and time as the court now becomes directly involved in case management. The Technology and Construction Court in the High Court has considerable experience of dealing with construction disputes. Before court proceedings are commenced, the parties should comply with the pre-action protocol for construction and engineering disputes. This requires the parties to set out their respective cases in correspondence and to meet on a without prejudice basis to seek to settle the dispute or narrow the issues involved, prior to the issue of court proceedings.

Mediation

Mediation is a voluntary and non-binding procedure. It is a private process in which an independent neutral person helps the parties reach a negotiated settlement.

Legislation and regulation

Adjudication under the Scheme for Construction Contracts - how to get started

Adjudication under the Housing Grants, Construction and Regeneration Act 1996 (the Construction Act) allows for a quick fix method of dispute resolution. The right to refer a dispute to adjudication is available to a party to a construction contract within the meaning of the Construction Act at any time.

Below is a brief explanation of the steps required to commence an adjudication under the Scheme for Construction Contracts, ie where there are no contractual adjudication provisions.

Where adjudication is your chosen method of dispute resolution, it is essential that you comply with the strict time limits laid down by the scheme and any timetable imposed by the adjudicator.

The adjudicator is under an obligation to reach a decision within 28 days of referral unless the parties agree otherwise. Therefore the responding party is not permitted to delay the proceedings without agreement. If no response is made to the adjudicator, he/she may issue an award on the basis of the evidence of the party which has requested the adjudication.

The decision of the adjudicator is binding on the parties until the dispute is finally determined by legal proceedings, arbitration (if the contract provides for arbitration or the parties agree to arbitration) or by agreement. Therefore, even if one party commences further proceedings, until the outcome of those proceedings is known, both parties will be bound by the adjudicator's decision and must comply with his award.

It is not possible to contract out of the time constraints laid down for adjudication by the Construction Act.

Procedure

To commence adjudication you must take three steps:-

- ❖ Give notice of adjudication.
- ❖ Request an adjudicator to act.
- ❖ Serve a Referral Notice.

Notice of Adjudication

This must be in writing, be given to every other party to the contract and contain:

- ❖ details of the parties involved;
- ❖ a brief description of the dispute;
- ❖ details of when and where the dispute arose;
- ❖ what you are seeking from the adjudicator, for example, an award for a specific sum; and
- ❖ names and addresses of the parties to the contract (including the addresses which the parties have specified for the giving of notices, if any).

Note: In view of the very tight timescale for adjudication you should ensure that your claim is fully prepared before issuing the Notice of Adjudication.

Legislation and regulation

Appointing an adjudicator

After giving notice of adjudication you must make a request for an adjudicator to act. The timescale for the appointment of an adjudicator is extremely tight and a request for the appointment of an adjudicator should be made at the same time as giving the Notice of Adjudication. In order to determine who to appoint you should consider the following:-

If your contract names an adjudicator, contact him/her to ensure that he/she is ready and willing to act.

If an Adjudicator Nominating Body (ANB) is named in your contract, contact that body and ask for an appointment to be made.

If no adjudicator or ANB has been named in your contract you can contact any ANB such as:-

The Academy of Independent Construction Adjudicators, 020 7608 5221.

The Royal Institution of Chartered Surveyors, 020 7222 7000.

The Chartered Institute of Arbitrators, 020 7837 4483.

The Royal Institute of British Architects, 020 7580 5533.

The Technology and Construction Solicitors Association, 020 7655 1000.

The Construction Industry Council, 020 7637 8692.

Contact the most appropriate ANB depending upon the nature of the dispute and the issues involved. Some ANBs will be able to offer a greater diversity and breadth of experience compared to other single discipline organisations.

If an ANB is used: The ANB has five days to inform you of the nominated Adjudicator. The nominated adjudicator then has up to two days to confirm his appointment.

Caution: If a named or nominated adjudicator refuses to act, another adjudicator can be agreed or nominated by any ANB, but beware the time limit for issuing the Referral Notice. If an alternative adjudicator cannot be appointed within seven days of the Notice of Adjudication the safest course is to issue a fresh Notice of Adjudication.

Referral Notice

Within seven days of the Notice of Adjudication you must send the Referral Notice to the adjudicator formally referring the dispute to him. The Referral Notice must be in writing, be given to the adjudicator and every other party to the dispute and:

- ❖ contain the basis of your claim, including an explanation of how the dispute arose and identifying the issues in dispute;
- ❖ be accompanied by copies of (or relevant extracts from) your contract (whether this is a standard printed form or evidenced in correspondence);
- ❖ include any documents upon which you wish to rely in support of your case;
- ❖ contain the remedies and award you are seeking; and
- ❖ give the adjudicator wide jurisdiction by giving him/her an alternative, for example, "such other sum as the adjudicator may determine".

It is important that the Referral Notice clearly sets out the issues in dispute which the adjudicator is being asked to determine including the history of the case and any arguments raised by the other party as well as identifying the remedies sought.

Following service of the Referral Notice the adjudicator should set a timetable for dealing with the adjudication including a response from the opposing party and request any further information or evidence in order to reach his decision.

Neighbourly matters

- Rights to light
- Daylight and sunlight
- Party wall procedure
- Access agreements
- Construction noise and vibration

Legislation and regulation

Legislation and regulation

Rights to light

Rights to light problems bring together two distinct but different areas of English law namely private nuisance – a sub-division of the law of torts; and easements – a sub-division of land law.

Private nuisance

The tort of private nuisance, like the tort of public nuisance, regulates activities affecting individual rights in or rights over real property (land).

A private nuisance may be defined as:

> **An unreasonable interference with a person's use or enjoyment of land itself, or some right over or in connection with land (ie a right to light).**

The law of nuisance tries to balance the legitimate activities of neighbours – a give and take approach.

Interference with a right to light must be objectively unreasonable in its extent and severity if it is to be sufficient to constitute a nuisance in the eyes of the courts.

Only when the courts are satisfied that the interference is unreasonable will they remedy the situation by awarding an injunction and/or damages.

It should be appreciated that often levels of natural light can be interfered with to a marginal extent and this will not necessarily constitute an infringement of a proprietary right that will be recognised as a nuisance.

Easements

An easement may be defined as a right annexed to land to use or to restrict use of neighbouring land in some way. More particularly, for a right to exist as an easement (as opposed to a restrictive covenant) the right must possess the four essentials:-

- ❖ There must be a dominant tenement and a servient tenement.
- ❖ The right must accommodate (benefit) the dominant tenement.
- ❖ The dominant tenement and the servient tenement must be owned or occupied by different persons.
- ❖ The right concerned must be capable of forming the subject matter of a grant.

Rights to light satisfy the aforementioned essentials and, like some other classes of right, have existed as easements for many centuries. The general rights to light principles set out below are distilled from the large body of case law that exists.

A right to light can be defined generally as:

> **A negative easement providing a right for a building to receive sufficient natural light through a defined aperture (usually a window) in perpetuity or for a term of years.**

Nature of a right to light

A right to light is not personal – it runs with property/buildings.

A right benefits the dominant tenement and burdens the servient tenement, but such a right will not necessarily be permanent. A right to light is for 'sufficient' natural light only and this is taken to mean enough light, "according to the ordinary notions of mankind" for:

Legislation and regulation

- comfortable use and enjoyment of a dwelling house; or
- beneficial use of and occupation of a warehouse, shop or other place (office, etc).

The test for 'sufficiency' is whether or not the dominant tenement will be left with enough light according to the ordinary requirements of mankind. Sufficiency is not based on the measure of light lost. See Colls v Home & Colonial Stores [l904] AC 179.

Actionable injury

No specific rule has been developed by the courts to define exactly when a reduction in natural light becomes actionable. The test for injury is uncertain but flexible. The court will have regard, in all cases, to the specific facts and circumstances, but, very generally, for day to day practical purposes, light specialists have adopted the general conventions that:

- A commercial property should be considered actionably damaged when less than 50% of an office floor area is lit to the critical one lumen (0.2% sky factor) level.
- A residential/domestic property should be considered actionably damaged when less than 55% of a room area is lit to the critical one lumen (0.2% sky factor) level.

It must be understood that the percentages mentioned above are not strongly founded in specific legal authority. The courts regard the percentages as good guides but not absolute tests.

Acquisition

A right to light may be created by:

- express grant or reservation (sometimes encountered);
- implied grant or reservation (rarely encountered); or
- prescription (very common and often called 'ancient lights').

Prescription means the procuring of a right on the basis of a long established custom and three methods of prescription exist, namely:

- time immemorial (right enjoyed since before 1189);
- doctrine of Lost Modern Grant (right enjoyed continuously for minimum 20 years); and
- Prescription Act 1832; Sections 3 and 4 (right enjoyed continuously for minimum 19 years and one day).

Prescription by the two above-mentioned common law doctrines can be defeated if it can be shown that the easement has not been enjoyed, 'as of right', for example, has not been enjoyed without force; without secrecy; without permission. In the City of London, because of the 'custom of London', the acquisition of a right by the doctrine of Lost Modern Grant is not available (see Bowring Services Ltd v Scottish Widows 1995).

Prescription, pursuant to the Prescription Act 1832, may be defeated if the servient party can show that:

- at some time within the last 19 years, they have prevented the entry of natural daylight through the subject apertures by erecting an opaque physical obstruction for a continuous period of at least one year;
- at some time in the last 19 years, they registered a 'national obstruction' pursuant to part 2 of the Rights to Light Act 1959; or
- the right has been enjoyed under some consent or agreement expressly given for that purpose by deed or in writing.

Legislation and regulation

Defence

A prescriptive right to light must be defended in the face of development.

Under section 4 of the Prescription Act 1832, if a party claiming a prescriptive right acquiesces in or submits to an interruption of light for one year or more, then their claim to a prescriptive right will be defeated. See Dance v Triplow and Another [1992] 17 EG, 103. Successful defence of a prescriptive right to light relies much on eternal vigilance and prompt protestation in writing to the obstructor. The protestations should also be repeated at regular intervals.

Remedies

The current legal system permits the awarding of:

- prohibitory or mandatory injunctions; and
- common law damages.

The courts are reluctant to permit a servient party to buy himself out of a breach of covenant and, generally, an injunction is regarded as the normal remedy, with damages the exception. The court may award damages in lieu of an injunction if all the four tests below can be answered in the affirmative:-

- Is the injury small?
- Would a small money payment be an adequate remedy?
- Would it be oppressive to the defendant to grant an injunction?
- Is the injury one that can be estimated in money terms?

These tests are derived from Shelfer v City of London Electric Lighting Co [1895] 1 Ch 287 31.

Compensation (ie damages) for injury to commercial property is usually valued on the basis of a 'freeholder in possession'. The 'freeholder in possession' compensation sum must be apportioned between the various interests in the dominant tenement using established valuation techniques.

With regard to residential property, the assessment of compensation is more subjective and it may be necessary to engage the services of a residential valuation expert local to the area concerned to offer opinion as to the diminution in value suffered.

Dos and don'ts

- Do establish the identity of all parties with an interest in the building and clarify exactly who you are acting for.
- Do consider whether the Crown or the local authority has ever had an interest in the servient tenement and whether the provisions of Section 237 of the Town and Country Planning Act 1990 apply.
- Do liaise closely with other specialists acting for other parties with injured interests in the dominant tenement.
- Do obtain copies of all leases and any restrictive/permissive deeds controlling rights to light.
- Do obtain copies (if available) of all up-to-date plans for both the dominant and servient sites/buildings.
- Do establish that a right to light exists in law.
- Do establish the extent of a right, ie number, size and location of apertures.
- Do consider whether transferred or 'incorporated' rights to light exist.
- Don't allow a 'dominant' party to be pressured into early agreement of compensation.

Legislation and regulation

- ❖ Do seek the advice of lawyers if the legal position is complicated beyond your experience by any controlling deeds or similar documents.
- ❖ Do ensure your client understands that the law relating to rights to light and the valuation techniques are not an exact science.
- ❖ Do explain to your client that there is no statutory procedure for dealing with rights to light issues and that there are no prescribed periods and deadlines during which or by which parties are obliged to settle issues.

Preventing the acquisition of a right to light

Problem

Under the Prescription Act 1832, the dominant owner will acquire a prescriptive right to light if the access to the light is enjoyed for 20 years without interruption (unless the right of access to the light has been enjoyed by written consent or agreement).

Prevention

To prevent the easement (in this case the right to light) being prescriptively acquired, the servient owner must:

- ❖ either obtain from the dominant owner agreement, in writing, to the effect that the access of light is actually enjoyed with the permission of the servient owner; or
- ❖ interrupt the enjoyment of the light for at least a year and with the knowledge of the dominant owner.

The latter may be done either by erecting a physical obstruction, or by registering a Light Obstruction Notice under Section 2(1) of the Rights of the Light Act 1959.

The light obstruction notice is registered with the local authority as a local land charge for a period of one year.

Procedure

The procedure to be adopted is as follows:

Stage 1. Obtain application forms from the Lands Tribunal (48-49 Chancery Lane, London WC2A 1JR. Tel: 020 7947 9200, Fax: 020 7947 7215 or download them from their website: http://www.courtservice.gov.uk/tribunals/ lands_frm.htm).

Complete Form 1 of the Lands Tribunal Rules 1996.

This is the application to the Lands Tribunal for a certificate to confirm that adequate publicity of the proposed application to the Local Authority has been given.

Complete Form A of the Local Land Charges Rules 1977.

*This is the application to the **local authority for the registration of the Light Obstruction Notice** as a local land charge.*

Return the forms to the Lands Tribunal with the fee and a coloured block plan indicating, in contrasting colours, the servient land, dominant land and the position and extent of the proposed opaque structure. (NB: this may be of unlimited height).

Stage 2. The Lands Tribunal will determine what publicity is to be given to the application and they will send out directions on how to, and who to, serve the notice on.

Stage 3. Once the notice is served in accordance with the directions given by the Lands Tribunal, a Certificate of Compliance should be sent back to the Lands Tribunal, which is effectively a letter confirming that the notice has been served on the persons instructed, together with evidence such as a recorded delivery slip or a letter from the person on whom notice was

Legislation and regulation

served (for example, the dominant owner) confirming receipt of the notice.

Stage 4. Once the Lands Tribunal are satisfied that the application has been publicised correctly, the Light Obstruction Notice will be registered with the local authority as a local land charge for the one year period, which represents the necessary period of interruption.

The applications are best made by the servient owner and/or his/her solicitor and the surveyor may advise as to the procedure to be followed and may assist with the production of the block plan, etc. If time is a critical factor (for example, the windows are just on the point of attaining their prescriptive right), you can ask for a temporary certificate as a matter of urgency, using the additional paragraph on Form 1. The Lands Tribunal will, if satisfied as to the necessity, grant registration for such period as may be specified in their temporary certificate (not exceeding six months) using Form 2, but then will extend this to 12 months with a definitive certificate when proof of publicity to the affected properties is produced. If you fail to comply with their directions for publicising the notice, or if the dominant owner proves that he/she already has a right to the light, then registration will lapse and no interruption will be deemed to have taken place.

Once a Light Obstruction Notice expires at the end of the one year period, the windows start to prescribe their right to light all over again. If the servient owner is to defeat the acquisition of a right to light in perpetuity, applications for Notional Obstructions must be submitted at maximum intervals of 19 years.

Further information

The two main statutes that are relevant to rights to light are:

- Prescription Act 1832; and
- Rights to Light Act 1959, Part II.

There exists a large body of rights to light case law. The selection of case reports listed below includes the more recent and more important decisions.

- Allen and Another v Greenwood and Another [1975] 1 All ER 819 6, 35-6
- Andrews v Waite [1907] 2 Ch 500 33
- Armitage v Palmer [1960] 175 EG, 315
- Attorney General v Manchester Corporation [1893] 2 Ch 87
- Bowring Services Ltd v Scottish Widows Fund & Life Assurance Society [1995] 16 EG, 206.
- Carr Saunders v Dick McNeil Associates Ltd and Others [1986] 1 WLR 992 37, 43
- Charles Semon & Co v Bradford Corporation [1922] 2 Ch 737
- Colls v Home & Colonial Stores [1904] AC 179 4, 9, 10, 29, 31-2, 34
- Cowper v Laidler [1903] 2 Ch 337, 341
- Dalton v Angus [1881] 6 App Cas 740
- Dance v Triplow and Another [1992] 17 EG, 103
- Davis v Marrable [1913] 2 Ch 421
- Deakins v Hookings [1994] 14 EG, 133
- Ecclesiastical Commissioners for England v Kino [1880] 14 ChD 213 40.
- Fishenden v Higgs and Hill Ltd [1935] 153 LT 128 33
- Gamble v Doyle [1971] 219 EG, 310 34
- Hawker v Tomalin [1969] 20 P & CR 550 10
- Hawkins v Rutter [1892] 1 QB 668 5
- Higgins v Betts [1905] 2 Ch 210

Legislation and regulation

- Leeds Industrial Co-Operative Society Ltd v Slack [1924] AC 851 45
- Lyme Valley Squash Club Ltd v Newcastle under Lyme Borough Council and Another [1985] 2 All ER 405 28-9
- Marine and General Mutual Life Assurance Society v St. James Real Estate Co Limited [1991]
- Mathias v Davies [1970] 214 EG, IIII 29
- Metaxides v Adamson [1971] 219 EG, 935 35
- Metropolitan Railway Co v Fowler [1892] 1 QB 165
- Moore v Hall [1878] 3 QBD 178 6. 43
- Newham and Others v Lawson and Others [1971] 22 P & CR 852 35
- News of the World Ltd v Allen Fairhead & Sons Ltd [1931] 2 Ch 402 32-3
- Ough v King [1967] 3 All ER 859 34
- Presland v Bingham [1889] ChD 268 40-1
- Price v Hilditch [1930] I Ch 500 5, 37, 43
- Pugh and Another v Howels and Another [1984] 48 P CR 298 36-7
- Scott v Pape [1886] 31 ChD 554 6, 80
- Sheffield Masonic Hall Co v Sheffield Corporation [1932] 2 Ch 17 43
- Shelfer v City of London Electric Lighting Co [1895] 1 Ch 287 31
- Smith v Baxter [1900] 2 Ch 138
- Tapling v Jones [1865] 20 CB (NS) 166 33, 40
- Wheaton v Maple [1890] 3 Ch 48 10
- Wheeldon v Burrows [1879] 12 ChD 31 78
- Williams Cory & Son Limited v City of London Real Property Company Limited [1954] 163 EG, 514 10, 33-4
- Wrotham Park Estate Co v Parkside Homes Limited [1973] ChD 321

Other References

'Rights of Light – The Modern Law' – S Bickford-Smith and A Francis

'Rights of Light' – J Anstey

'Easements, Rights to Light' – J P Dooley and M Gunson, Estates Gazette

Daylight and sunlight

As greater emphasis is placed on environmental issues, local planning authorities are increasingly concerning themselves with the effect of developments on the daylight and sunlight enjoyed by neighbouring properties. The local authority's Unitary Development Plan should always be consulted, as it will give an indication of what they expect in this regard.

The Building Research Establishment (BRE) published Report 209 in 1991 titled 'Site layout planning for daylight and sunlight: A guide to good practice' written by P. J. Littlefair. It was intended to give guidance on how to avoid detrimentally affecting neighbours' daylight and sunlight and how to ensure good daylight and sunlight to proposed new development through good design. It sets out various tests that can be undertaken to establish if a problem is likely to be created.

The BRE report was not intended to be mandatory. However, it is not uncommon for the planning authority to require a developer to submit a daylighting and sunlighting study in support of a planning application and some may expect compliance with the guidelines.

Legislation and regulation

Daylight

The BRE report states that if any part of a new building or extension, measured in a vertical section perpendicular to a main window wall of an existing building, from the centre of the lowest window, subtends an angle of more than 25° to the horizontal, then the diffuse daylighting of the existing building may be adversely affected. This will be the case if either:

- the vertical sky component measured at the centre of an existing main window is less than 27%, and less than 0.8 times its former value; or
- the area of the working plane in a room which can receive direct skylight is reduced to less than 0.8 times its former value.

In such circumstances the occupants of the existing building will notice the reduction in the amount of skylight and more of the room will appear poorly lit.

Sunlight

The BRE report advises that new development should take care to safeguard access to sunlight for existing dwellings and any non-domestic buildings where there is a particular requirement for sunlight.

If a living room of an existing dwelling has a main window facing within 90° of due south, and any part of a new development subtends an angle of more than 25° to the horizontal measured from a point 2 metres above ground in a vertical section, perpendicular to the window, then the sunlighting of the existing dwelling may be adversely affected.

The sunlighting of the existing dwelling will be considered to be adversely affected if the window reference point:

- receives less than one quarter of annual probable sunlight hours, and/or less than 5% of annual probable sunlight hours during the winter months between 21 September and 21 March, and the available sunlight hours in either period is reduced to less than 0.8 times its former value.

Analysis

The BRE report sets out the methods by which the effect of a development on existing neighbouring buildings may be calculated. If the local planning authority requires an assessment to be undertaken and submitted in support of a planning application it is usual for the analysis to follow the tests in BRE Report 209.

Party wall procedure

In England and Wales, party wall matters came into force on 1 July 1997 under the jurisdiction of the Party Wall etc Act 1996.

All previous local enactments, including the London Building Acts (Amendment) Act 1939 Part VI and the Bristol Improvement Act 1847, have now been repealed.

As implemented, the Act only covers the whole of England and Wales. While it is not intended that it will ever cover Scotland, it will probably eventually become law in Northern Ireland.

The general principle of the Act is to enable an owner to undertake works on, or adjacent to, adjoining properties while giving protection to potentially affected neighbours.

The definition of a party wall is one that stands on the land of two owners, by more than its footings, or one which separates buildings of different

Legislation and regulation

owners. In the first case, the whole wall is a party wall whereas in the second, it is only a party wall for the extent to which the two buildings are using it and the whole of the rest of the wall belongs to the person on whose land it stands.

There can be many owners of such walls including freeholder, leaseholder and anyone with an interest greater than from year to year. Under the Act, the word 'owner' can also include people with a contract to purchase or an agreement for lease; this arrangement allows an intending owner to serve notice, and even commence work, before completion of the contract.

Each owner owns the part of the wall that stands on their own land, but also has rights over the remainder of the wall. The Act allows owners to treat the whole of the party wall as if it were their own, and debars them from dealing with their half on its own without informing their neighbours.

Before exercising any of the rights bestowed upon them, owners must follow the procedures set down in the Party Wall etc Act 1996. (The 'etc' in the title refers to other works, such as adjacent excavations and building walls at the boundary).

If an owner wishes to raise, cut into, thicken or demolish and rebuild a party wall, they must give notice of their intention to do so. In the event that the adjoining owner disagrees, each party must appoint a surveyor and a formal agreement known as a party wall award must be entered into. The same is true in relation to deep foundations or excavations within prescribed proximity to adjacent buildings.

Refer to figure below for details of excavations within 3 metres and 6 metres that fall under the Act and require notice to be served.

3 metre notice

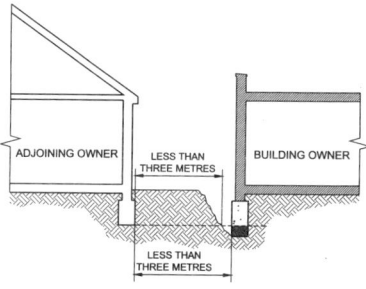

6 metre notice

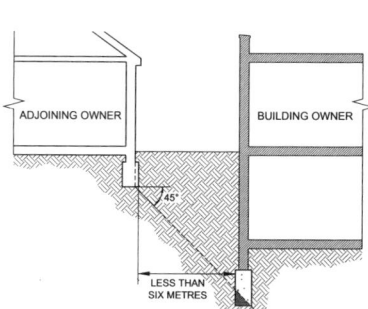

Legislation and regulation

Definitions and procedures are set down in the Party Wall etc Act 1996 and some of the main sections included are as follows:-

- Section 1 deals with the construction of a wall on a boundary where no other construction currently exists.
- Section 2 gives a building owner the right to undertake certain works to a party structure and certain limited works to an adjoining owner's independent structure.
- Section 3 confirms the arrangements for serving Party Structure Notices at least two months in advance.
- Section 4 confirms the arrangements for serving counter notices. These should be served within one month.
- Section 6 confirms the arrangements for notices in regard to foundations or excavations within 3 metres, or within 6 metres and a prescribed angle.

Remaining sections of the Act clarify the rules and procedures under which the Act is administered. They also clarify a few sets of buildings that are exempt from the regulations.

Dos and don'ts

- Do remember that the party wall surveyor is administering the Act impartially and not representing a client.
- Do establish with certainty the adjoining owners' interests.
- Don't be affected by conflict of interest – remain impartial.
- Do clarify extent of work to affect the party structure.
- Do ensure continuation of interest and cooperation of all parties including the design team and contractor.
- Do obtain requisite information for same in good time to obtain agreements.
- Do be prepared to negotiate terms of the award – do not stand fast.
- Do respond to all correspondence and information sent immediately – this will save time and confusion in the future.
- Do agree awards prior to the commencement of work.
- Don't confuse common law and rights to light matters as being part of party wall procedures.
- Do establish duties of surveyors.
- Don't allow adjoining owners to be antagonised during the work – 'keep the peace'.
- Do establish a line of communication between adjoining owners and contractors for airing grievances.
- Do ensure that all parties fulfil their obligations.
- Do allow for sufficient time between service of notice, and the start of works.

References

'Party wall legislation and procedure' RICS guidance note (fifth edition) published by Surveyors Publications.

'The Party Wall Act explained – a commentary on The Party Wall etc Act 1996'- published by The Pyramus and Thisbe Club.

'Party Walls – the new law' by S Bickford-Smith and C Sidenham, published by Jordans.

'Practical manual for party wall surveyors' by J Anstey, published by RICS Books.

'Party walls and what to do with them' by J Anstey, fifth edition, published by RICS Books.

'Party Wall etc Act 1996' audio CD, published by Owlion.

Legislation and regulation

The Party Wall etc Act 1996 is available on-line at the HMSO website: http://www.legislation.hmso.gov.uk

Access agreements

Access to Neighbouring Land Act 1992

Often confused with the Party Wall etc Act 1996, the Access to Neighbouring Land Act 1992 enables a court to grant an order for access to land where such access is required to enable the execution of repair or maintenance work and where access has been refused by the neighbour.

The Act contains provisions for the preparation of schedules of condition and for overseeing of the work by surveyors. Surveyors can also be called upon to provide evidence where claims for damage under the Act are made.

Often the knowledge that rights exist under the Act will prompt neighbours to be accommodating, but, even then, it is still sensible to enter into an informal access agreement in the form of a licence with accompanying schedule of condition.

Rights of way and fire escape agreements

More intense use of an existing right of way often requires re-negotiation of the right altogether. One type of right commonly encountered relates to fire escape routes benefiting properties that adjoin a development. The rights are often disrupted by large-scale redevelopment and negotiations for both temporary escape rights during the redevelopment and for revised escape rights over the new permanent development will be required.

Oversailing cranes and encroaching scaffolds

Most large developments necessitate the use of at least one crane. There are benefits to the developer or contractor of using fixed-jib tower cranes, as opposed to luffing-jib or folding-jib cranes, such as reduced cost and increased speed and load capacity. However, where the jib of a crane will oversail adjoining land, the agreement of the adjoining land owner, and any other party with an interest in the air space above it, will be needed. Without such agreement the developer will be committing a trespass on every oversailing occasion.

The law on this matter comes from the tort of trespass. The position was clarified in the case of Anchor Brewhouse Developments Ltd and others v Berkeley House (Docklands Developments) Ltd [1987], in which the plaintiffs sought, and were granted, injunctive relief in relation to an unauthorised oversailing crane.

While some developers manage to place oversailing risks on their contractors, when work is abundant such contractors will often not accept such risks or, if they do, their tender prices are significantly enhanced.

Similar trespass issues often arise in relation to temporary independent scaffolding, although, wherever possible, rights granted by the Party Wall etc Act 1996 and the Access to Neighbouring Land Act 1992 should be exercised.

Developers and/or their contractors should give consideration at an early stage to the issue of access onto or over adjoining land or the airspace above it. Neighbours should be approached in advance for consent and the terms of such consent will be the subject of negotiation by the parties. It is sensible to enter into an access agreement or licence and the developer may be expected to pay a sum of money in consideration of grant reciprocal oversailing rights. Typically such agreements will provide for indemnities, insurance arrangements, schedules of condition, payment of fees and costs and so on.

Legislation and regulation

Construction noise and vibration

The law relating to construction noise and vibration is controlled by the common law of nuisance and a number of statutes, in particular, the Control of Pollution Act 1974.

Nuisance

A private nuisance is "an unreasonable interference with a person's use or enjoyment of the land". If a neighbour can demonstrate that a nuisance exists and that he has suffered substantial damage as a result, then he may be successful in bringing an action against the parties creating the nuisance.

The remedies available are injunctions and/or damages. The courts have, however, traditionally regarded demolition and construction sites as a special case as far as noise is concerned. It would appear that as long as the works are carried out with proper skill and care, and all reasonable precautions are taken to minimise disturbance to neighbouring occupiers, no action in nuisance will arise.

Developers and contractors have to ensure that they take "reasonable precautions". They must be aware of the definition of "best practicable means" in the Environmental Protection Act 1990. This includes considering local conditions and circumstances, the current state of technical knowledge and financial implications.

A leading case on construction noise and nuisance is Andreae v Selfridge [1938] 3 All ER 264 where the judge held that provided demolition and building operations are "reasonably carried on, and all proper and reasonable steps are taken to ensure that no undue inconvenience is caused to neighbours, whether from noise, dust or other reasons, the neighbours must put up with it".

Control of Pollution Act 1974

The Control of Pollution Act 1974 deals, in part, with the control of noise on construction sites. Section 60 empowers a local authority to serve notice imposing certain limitations. These limitations include specifying the hours of work, permitted noise levels and the particular plant and machinery that may be used. The recipient may appeal against the notice within 21 days. Contravention of the notice is an offence under the Act.

Section 61 provides an opportunity for a developer or contractor to apply to the local authority in advance of the works and seek agreement to such matters as the method of carrying out the work and the steps that will be taken to minimise noise. The local authority is not obliged to give its consent and, even if it does, it may attach conditions to it. It may also change the conditions if it sees fit.

The local authority will often specify that "best practicable means" are used. The developer/contractor will be expected to adopt the quietest viable method of working within reasonable cost limits. The local authority may also require continuous noise and vibration monitoring and regular liaison meetings with neighbours.

Architectural and design criteria

Architectural, engineering and services design

- Basic design data
- Building types
- Specifications
- Different specifications for different contracts
- Specification writing
- Site archaeology
- Coordinating project information

Architectural, engineering and services design

Basic design data

Metric conversion for commercial property areas

- To convert square feet to square metres x factor 0.0929.
- To convert square metres to square feet x factor 10.764.

Internal circulation

Space allowances: (minimum areas per person)

Building type	m^2	ft^2
Offices (excluding cores)	9.3	100
Retail	4.6-7.0	50-75
Factories	7.0	75
Restaurants	0.9-1.1	10-12

Lighting requirements

Circulation: 150 Lux
Casual work: 200 Lux
Routine work: 300 Lux
Offices: 500 Lux
Drawing office: 750 Lux
Fine work: 1000 Lux
Very fine work: 1500 Lux
Minute work: 3500 Lux

Noise and acoustics requirements

Auditoria: 20-30 dB (A) Leq
Bedroom: 30-35 dB (A) Leq
Small office: 40-45 dB (A) Leq
Large office: 45-50 dB (A) Leq
Light industrial: 50-55 dB (A) Leq

External circulation

Car parking	Standard bay 2.4 x 4.8m	Allow 6.1m for head-on parking. Area per car 18.8m^2
Ramps	Car parking garages	10%
	With transition ramps at half the ramp gradient for 2.4m at each end	30%
	Pedestrian	10%
Carriageway widths	One-way four lanes	14.6m
	One-way two lanes	7.3m
	Two-way two lanes	max 7.3m, min 6.0m
Vehicle sizes	Cars max length	5.7m
	Cars min length	3.05m
	Lorries max length	18.0m
	Vans max length	6.0m
	Standard refuse lorry length	7.4m
	Standard fire appliance length	8.0m

Architectural, engineering and services design

Building types

Housing

Typical approximate housing equivalent densities:

```
50 houses per acre = 124 houses per hectare
40 houses per acre = 99 houses per hectare
30 houses per acre = 74 houses per hectare
20 houses per acre = 49 houses per hectare
10 houses per acre = 25 houses per hectare
```

Recommended planning grid – 300mm
Recommended minimum floor to ceiling – 2300mm
Minimum floor to floor – 2600mm
Recommended areas for number of persons per dwelling (m²):

Houses:	1	2	3	4	5	6	7
One storey	30	44.5	57	67	75.5	84	
Two storey				72	82	92.5	108
Three storey					94	98	112
Flats	30	44.5	57	70	79	86.5	
Maisonettes				72	82	92.5	

Parker Morris is the preferred standard for new buildings by local authorities and housing associations. Private housing is subject only to space requirements by the Public Health Acts.

Hotels

Types
- city centre hotels
- motor hotels
- airport hotels
- resort hotels
- motels

Space allocation (m²/room gross)
- city centre hotel   45–65
- motor hotel   35–45
- resort hotel   40–55

Hotels under 77–80 rooms only viable as family run businesses.

Fire precautions
Travel distances for escape are dealt with in the guide to the Fire Precautions Act 1971.

Car parking
- resident guests: one space/bedroom
- conference facilities: one space/five seats

Required provision must be agreed with the local authority.

Offices

Methods of calculating areas
- For planning purposes, gross total area measured over external walls.
- For cost purposes, gross total area measured inside external walls.
- Nett areas are measured between inside walls and exclude core areas, ducts and staircases.

Architectural, engineering and services design

Definition of space
- very deep space: over 20m
- deep space: 11-19m
- medium deep space: 6-10m
- shallow space: 4-5m

Dimensional criteria
- planning grids: 900mm 1200mm 1500mm
- floor-to-floor heights: 2700mm-5100mm
- floor-to-ceiling heights: 2400mm-3000mm
- floor zone: 300mm-1200mm

Means of escape
- Maximum travel distance with escape in only one direction – 12.2m.
- Maximum travel distance with escape possible in alternative direction – 46m.
- Maximum distance between exits on a storey – 61m.

At least one fire fighting stair required with floor more than 18.3m above ground floor level.

Lavatory provision based on the Offices, Shops & Railway Premises Act 1963 Sections 9 & 10.

Car Parking
- Staff: one space for each 25m² of gross floor area.
- Visitors: 10% of parking provision required.

Required provision must be agreed with the local authority.

Recreational

Sports facilities outdoors:

Swimming pools. Olympic standard:	length width constant depth	50m 21m 1.8m
Running tracks:	200m 300m 400m	86.79 x 61.78m 126.52 x 86.48m 170.91 x 133.00m
Association Football pitch size		100m x 60m
Rugby Union pitch size		144m x 69m
Hockey pitch size		91.5m x 54.9m
Tennis court size		0.97m x 23.77m
Sports facilities indoors: Netball court size		30.50m x 15.25m
Badminton court size		13.4m x 6.1m
Table tennis (international)		14.0m x 7.0m
An indoor space 32m x 26m can contain the following sports:	Two badminton courts karate or fencing or table tennis basketball or archery or volleyball or soccer or hockey	

Architectural, engineering and services design

Industrial

Site coverage:	Plot ratio normally maximum of 1:1 including office content. Site coverage should not exceed 75% (normally approx. 50-60%)	
Car parking: Factories	Staff:	1 car/50m² of gross floor area
	Visitors:	10% of staff parking
Warehouses:	Staff:	1 car/200m² of gross floor area.
	Lorry parking: Minimum standards for loading bays:	70m² for every 100m² gross floor space
		140m² for every 250m² gross floor space
		170m² for every 500m² gross floor space
		200m² for every 1000m² gross floor space
		300m² for every 2000m² gross floor space
		50m² for every additional 1000m² gross floor space
Factory building types:	Light duty industrial:	Spans min 9m max 12m Ht. to eaves 4.5m floor loading 16kN/m²
	Medium duty industrial:	Spans 12m-18m Ht. to eaves 6.5m Floor loading 25kN/m²
	Heavy duty industrial:	Spans 12m-20m Ht. to eaves 7m-12m Floor loading 15-30kN/m²
Warehouses:	General purpose:	Spans 12-18m Ht. to eaves 8m Floor loading 25kN/m²
	Intermediate high bay:	Spans 12-20m Ht. to eaves 14m Floor loading 50kN/m²
	High bay:	Ht. to eaves 30m Floor loading 60kN/m²

Architectural, engineering and services design

Specifications

The specification is the hub around which the various documents forming a modern day building contract are assembled.

The specification will incorporate a complex array of information, which will include design, testing, procedural, visual and quality control information.

The parties will seek to have fully prescribed the product to which they are to sign up under a building contract. The standard forms of building contract generally recognise the specification as a contract document, but there are exceptions such as the JCT98 Standard Form with Quantities, which does not and care must be taken to ensure that the wording is amended to include the specification.

The specification as a building contract document will typically be supported by the preliminaries and either bills of quantities or contract sum analysis. It will schedule the drawings to which it refers, which should form the contract drawings.

Different specifications for different contracts

Design specification

This form of specification will be used where the author wishes to prescribe the project in as much detail as possible, no imagination is required as to the clients' requirements. Products will be named and the assembly of the building will be described and supported with drawn information on a 'do as I say' basis.

The form of contract employed here will be a traditional one which will not carry a design responsibility for the contractor.

Performance specification

A performance specification sets out design parameters to which contractors are invited to submit their proposals. This can be a procurement route used where the design team have little idea as to products or visual requirements.

The client will not need to advance the design as far before inviting tenders from interested contractors. This is a more economic approach since pre-contract design costs are usually less at the early stages of procurement.

The specification for a design and build contract will initially form the employers requirements latterly supported by the contractors proposals once agreed with the selected contractor. This approach has the benefit of the contractor bringing the supply chain knowledge to the client at an early stage while still in competition.

The parameters of design so described in a performance specification can either be laid down loosely or brought in very tight according to the degree of control the client wishes to retain.

Specification writing

Construction consultants have established standardised data at their disposal to assist and form the basis to prepare their documents. Online specification writers are available, the most widely used is National Building Specification (NBS) - www.nbsservices.co.uk. This is a subscription service.

Architectural, engineering and services design

The NBS system stores a fully comprehensive set of specifications for most conceivable products and materials which they may adapt, add to or modify according to a particular project.

The NBS has recently published 'NBS Educator' which is designed to provide construction industry students, lecturers and professionals who need to understand contract documentation, such as specifications and schedules of work. The NBS Educator website includes a variety of functions to navigate which provides the user with suggestions and ideas as well a reference library for different aspects of specification writing.

Another specification tool that can be accessed online is Barbour Index which has recently launched its Specification Expert website which is at www.barbour-index.co.uk. The service is designed to create, edit and manage specification documents. The service makes writing specifications quick and easy, offering a comprehensive library of all major specification clauses. It is available in two versions: Building (Architectural and, Building Surveying) and Building Services which incorporates the National Engineering Specification.

Site archaeology

The 'Little Oxford Dictionary' definition of archaeology is:

> Archaeology - the study of ancient peoples, especially by excavation of physical remains.

Not everyone will realise that in the construction world we have to deal with two types of archaeology:-

- ❖ The traditional excavation of what lies hidden below ground.
- ❖ The archaeology of standing buildings or above-ground archaeology.

How does archaeology affect construction?

Archaeology affects construction primarily through the way that the planning process limits and controls development. Both above and below ground archaeology can easily be applicable to the same site.

The present key government reference document is Planning Policy Guidance Document 16 published in November 1990 (PPG16). The government is now consulting for the replacement for PPG16 with the first of the new Planning Policy Statements (PPS). In this case the Planning Policy Statement on Planning for the Historic Environment.

There is at present no formal indications as to when PPG16 is to be replaced by the PPS although at the time of writing (mid November 2002) the draft is anticipated by early December 2002.

PPG16 sets out the secretary of state's policy on archaeological remains on land, and how they should be preserved or recorded both in an urban setting and in the countryside.

A key statement is paragraph 6 of PPG16 which quoted in full states:

> Archaeological remains should be seen as a finite, and non renewable resource, in many cases highly fragile and vulnerable to damage and destruction. Appropriate management is therefore essential to ensure that they survive in good condition. In particular, care must be taken to ensure that archaeological remains are not needlessly or thoughtlessly destroyed. They can contain irreplaceable information about our past and the potential for the increase in future knowledge. They are part of our sense of national identity and valuable both for their own sake and for their role in education, leisure and tourism.

Architectural, engineering and services design

And an extract from paragraph eight of PPG16:

> Where national important archaeological remains, whether scheduled or not, and their settings, are affected by proposed development there should be a presumption in favour of a physical preservation.

Planning permission is required for works of development – the carrying out of building, engineering, mining or other operations in, on over or under land or the making of any material change in the use of any buildings or other land.

The first port of call for any planning application is the local planning office and the published documentation in either the Development Plan or the Unitary Development Plan.

Paragraph 15 of PPG16 is respect of Development Plans states:

> Development Plans should reconcile the need for development with the interests of conservation including archaeology. Detailed Development Plans (ie Local Plans and Unitary Development Plans) should include policies for the protection, enhancement and preservation of sites of archaeological interest and of their settings. The proposals map should define the areas and sites to which the policies and proposals apply. These policies will provide an important part of the framework for the consideration of individual proposals and development which affect archaeological remains and they will help guide developers preparing planning applications.

In paragraph 16, of PPG16 in respect of archaeological remains of National Importance scheduled as ancient monuments under the Ancient Monuments and Archaeological Areas Act 1979 we find:

> Archaeological remains identified and scheduled as being of National Importance should normally be earmarked in development plans for preservation. Authorities should bear in mind that not all nationally important remains meriting preservation will necessary be scheduled; such remains and, in appropriate circumstances, other unscheduled archaeological remains of more local importance, may also be identified in development plans as particularly worthy of preservation.

Also, all counties maintain a Sites and Monuments Record of both scheduled ancient monuments and most importantly also locations **"where archaeological remains are known or are thought likely to exist"**.

If the Local Development Plan and/or the Local Planning Office identifies the site as one with scheduled ancient monuments or locations where archaeological remains are believed to exist then the earliest possible consultation with the county archaeologists or their London equivalents at English Heritage is vital.

This consultation will identify the archaeological sensitivity of the site.

Following discussions with the county archaeologists the next step for the developer is to have researched exactly what is or might be there and how it will or could impinge on his proposed development.

This is a cost for the developer and will have to be paid for by the developer in the same way as any other site investigation exercise is paid for prior to a development being undertaken.

Options for archaeological investigations are potentially twofold.

Desktop studies

These draw on the archaeological records of the site itself and those of neighbouring sites. They are relatively quick and easy. Crucially they must be done by a trained archaeologist. The handbook of the Institute of Field

Architectural, engineering and services design

Archaeologists will help in locating the right archaeologist or archaeological unit to undertake the study. Contact can be made on admin@archaeologists.net or by telephone on 01189 316446.

Discussions may however indicate that the next level of investigation is necessary.

Field evaluation

This is not a full archaeological elevation or 'dig' but a ground survey and small scale trial trenching.

The results from the desktop study and/or the field evaluation will inform both the developer and the county archaeologist and the local planning officers, especially the conservation officer, as to the importance of the site in archaeological terms.

PPG16 in paragraph 22 states:

> Local Planning Authorities can expect developers to provide the results of such assessments and evaluations as part of their application for sites where there is good reason to believe there are remains of archaeological importance. If developers are not prepared to do this voluntarily the planning authority may wish to consider whether it would be appropriate to direct the applicants to supply further information and, if necessary, authorities will need to consider refusing permission for proposals which are inadequately documented.

PPG16 also states in paragraph 25:

> Planning authorities should not include in their development plans policies requiring developers to finance archaeological works in return for the grant of planning permission.

and

> Where planning authorities decide that the physical preservation in situ of archaeological remains is not justified in the circumstances of the case and that development resulting in the destruction of the archaeological remains should proceed, it would be entirely reasonable for the planning authority to satisfy itself before granting planning permission, that the developer has made appropriate and satisfactory provision for the excavation and recording of the remains. Such excavation and recording should be carried out before development commences, working to a project brief prepared by the planning authority and taking advice from the archaeological consultants. This can be achieved through agreements reached between the developer, the archaeologist and the planning authority. Such agreements should also provide for the subsequent publication of the results of the excavation. In the absence of such agreements, planning authorities can secure excavation recording by imposing conditions on the planning permission.

Model agreements between developers and the appropriate archaeological body regulating archaeological site investigations and excavations can be obtained from the British Property Federation. **The email address is publications@bpf.org.uk or by telephone on 020 7828 0111.**

Statutory protection for archaeological sites

Where Scheduled Ancient Monuments have been identified under the Ancient Monuments and Archaeological Areas Act 1979, these are sites of national importance and as such rank as the equivalent of Grade I or Grade II* buildings of special architectural or historic interest.

The protection of Scheduled Ancient Monuments is based on the

Architectural, engineering and services design

requirement that scheduled monument consent has to be obtained from the secretary of state (now of the Department of the Environment and the Regions, previously the Department of the Environment).

Application has to be made to the secretary of state for any works that would have the effect of:

> Demolishing, destroying, damaging, removing, repairing, altering, adding to, flooding or covering up the monument.
>
> Since Scheduled Ancient Monuments include both below ground monuments and above ground monuments Scheduled (Ancient) Monument consent can have a very real affect on the plans of a site owner. If you find a scheduled ancient monument standing on or laying below your site take great care. Ensure your professional advisors fully appreciate the implications before application is made. If in doubt, enter into the earliest possible discussions with the relevant heritage and conservation authorities covering the site. These will include the local authorities conservation officer, the county archaeologist, and the Inspectors of Ancient Monuments at English Heritage.

Above ground archaeology

A Scheduled Ancient Monument is usually, although not exclusively, an unoccupied structure and can be either remains below ground and/or a standing structure above ground.

Above ground Scheduled Ancient Monuments will require Scheduled Ancient Monuments Consent for works to the structure. To fully understand what is important in this structure is understanding 'ancient peoples by the study of their physical remains' was borne the concept of above ground archaeology. This can involve the physical digging in to the structure above ground. This can equally involve the 'desktop study' approach of bringing together the known archive or history of the structure to fully understand what is important in the remains and what is less important.

This is because as structures are added to and/or changed over time there may well be pieces or parts of what is now extant that can well be removed or modified or changed and benefit the scheduled monument.

The final use of archaeology in standing structures is the extrapolation of the 'desktop study' for above ground scheduled ancient monuments into the 'Conservation Plan' for all types of listed historic buildings.

Conservation Plans are increasingly required by the Heritage and Listed Building authorities, particularly for the more important Grade I and Grade II* buildings as listed as being of Special Architectural and Historic Interest. This is as part of the application process for Listed Building Consent for works that will affect the special architectural or historic interest of these Listed Buildings.

Conservation plans have a number of practical and financial advantages for the developer/renovator or historic buildings. The key ones are:

(a) Certainty for the design team in formulating in what can or cannot be done to the building and site before they formulate details for the Listed Building Consent and Planning Permission applications;

(b) Certainty for the developer/renovator that the scheme will not be stopped in its tracks by the Heritage authorities discovering that the building contains important previously unknown historic elements in the structure that precludes all or part of the development going forward. The worst case being the stopping of the scheme or at least making it less viable or completely unviable.

Architectural, engineering and services design

Coordinating project information

Some of the characteristics of the construction industry make the quantity and timing of information flows on construction projects difficult to administer. These include the following:-

- Large numbers of participating companies collaborating on projects of relatively short duration.
- Lack of control over the location and working conditions of construction sites.
- Comparatively low levels of management support.

Despite this, use of electronic systems for information management including electronic tendering and project management systems claim to have several benefits including:

- reduced paper trails on labour intensive tasks;
- reduced costs and time savings;
- Improved management of information;
- a platform to stimulate innovation; and
- the support of a green agenda.

Relevant legislation/information

The British Standards Institution's section for delivering information solutions to customers (BSI-DISC) sets out five principles which should be the starting point for coordinating information in 'DISC PD 0010:1997, The principles of good practice for information management' as follows:-

- Recognise and understand all types of information.
- Understand the legal issues and execute 'duty of care' responsibilities.
- Identify and specify business processes and procedures.
- Identify enabling technologies to support business processes and procedures.
- Monitor and audit business processes and procedures.

There is a need at the inception of any construction project to agree the principles in order to effectively manage communication and the flow of information. The above principles are relevant whether a web-based system is being used on a large prime contract project or a paper-based system on a smaller traditional project. The principles should be agreed with all members of the project team and documented with their conditions of appointment.

The DTI, Building Centre Trust and Phrontis Ltd have in partnership produced a Construction Best Practice report titled 'Effective Integration of IT in Construction' (October 2001). It highlights three key sectors in which e-business can offer improved means of coordinating project information including; specification, tendering and the supply chain.

- The internet is currently providing many opportunities for materials and equipment firms to provide information and communicate more directly with specifiers.
- In connection with e-tendering, the Office of Government Commerce (OGC) has been using a pilot system for several contracts and it is the governments intention to introduce e-tendering in central civil government by the end of 2002.
- To achieve success for e-business supply chains is difficult. The nature of relationships is generally transitory, driven by project type and locality. Contractors are generally put off investing in this area because of the threat of reduced margins.

The report concludes that e-business offers the construction industry a powerful array of tools to communicate and collaborate and should go some way to help overcome many of the endemic problems of the industry.

Architectural, engineering and services design

'A Code of Procedure for Construction Industry on Production Information' has been issued for comment by the construction industry (issued by the DTI/CCPI September 2002). The aim of the code is to provide authoritative practical guidance on the preparation of drawings, specifications and schedules of work making optimum use of widely adopted systems.

'BS7799 Information Security Management' provides guidance on security of information, which is of particular relevance within partnering relationships. A continuing push to use IT to conduct business electronically and globally requires a high degree of trust between customers and suppliers and between trading partners. The aim of BS7799 is to give guidance on internal management systems, to improve confidence between trading partners and to ensure compliance with the 1998 Data Protection Act.

A report entitled 'Benchmarking Electronic Service Delivery' by the central IT unit of the Cabinet Office compares the UK's progress on the development of electronic government services with that of other major economic nations. The UK has set more ambitious targets than most of its competitors with an aim to provide 100% of all government services electronically by 2005.

Examples of project management systems

The following list summarises a small sample of project management systems currently available on the World Wide Web:

Asite – www.asite.com

Cadweb – www.cadweb.co.uk

Citadon – www.citadon.com

Meridian – www.mps.com

Unisys – www.unisys.co.uk

Recommendations

The recommendations for overcoming barriers to IT integration for coordinating project information, as summarised from 'Effective Integration of IT in Construction' include the following:-

- ❖ Adopt a suitable legal framework for using electronic communications.
- ❖ Appoint a leader with IT skills early in a project.
- ❖ Develop an information management strategy at project inception.
- ❖ Consider integration skills as a factor in the selection of the professional team.
- ❖ Ensure each team participant's in-house information systems are robust.
- ❖ Consider using a collaborative approach.

In conclusion, a complete understanding of the project, the environment and the people involved is necessary to effectively coordinate project information.

Structural and civil engineering design

- The design process
- Soils and foundation design
- Soil survey
- Design loadings for buildings

中
Architectural, engineering and services design

The design process

The Institution of Structural Engineers defines structural engineering as:

> The science and art of designing and making, with economy and elegance, buildings, bridges, frameworks and similar structures so that they can safely resist the forces to which they may be subjected.

Generally, engineers use limit state design methods. These model the way in which loadings are applied to a building over the course of its life.

Limit state design utilises differing probability factors for separate types of loads and material stresses to reflect the degree of certainty in assumptions made about each. For instance, strength is checked at the ultimate limit state where the building or element is on the point of collapse.

Deflection and vibration are checked at the serviceability limit state, at which point the building is considered to be unserviceable, even though it will not collapse.

Today, design in concrete, steel and masonry is usually carried out using limit state methods, with timber design scheduled to change over with the imminent arrival of the new Eurocode. Foundation design, involving the much more unpredictable material of the soil, has still eluded the process.

An assessment of ground conditions is usually essential to ensure the proper performance of the superstructure.

Soils and foundation design

For very light residential building, a visual assessment of the soils with a minimum of physical testing may be sufficient, if local knowledge of the area is good. For larger buildings, a proper geotechnical investigation is usually recommended. This allows the measurement of relevant soils (geotechnical) data on which the foundation design will be based. Accurate testing of samples of soil obtained from boreholes usually leads to economies in foundation design. Samples are usually tested for consolidation, compressibility (both of these affect settlement) and shear strengths. In certain circumstances testing for chemicals, which could affect the structure, may be included.

Where soil conditions are good, and building loads are relatively light, a simple spread foundation is usually sufficient. This will be in the form either of simple strips or pad bases of non-reinforced concrete. As loads increase and soil conditions become more complex, foundations may need to be reinforced to allow loads to be spread sufficiently. Heavy building loads and poor ground often require piled foundations to take the foundation loads into suitable (deeper) soils.

Other features, such as trees (in shrinkable soils), groundwater, and basements, all require special consideration in foundation design.

The relevant British Standards to which reference should be made are BS 8004 Code of Practice for Foundations and BS 5930 Code of Practice for Site Investigation.

For preliminary design purposes only, the following table gives typical allowable bearing values under static loading for various types of soils and may be used to size strip or pad foundations.

Architectural, engineering and services design

Category	Soil type	Presumed allowable bearing pressure (kN/m^2)
Un-weathered rocks	Strong igneous rocks,	10,000
	Strong sandstones and limestones	4,000
	Strong shales	2,000
Non-cohesive soils	Dense gravel	600
	Medium dense gravel or sand and gravel	200-500
	Loose gravel or loose sand and gravel	up to 200
	Loose sand	up to 100
Cohesive soils	Very stiff clay	300-600
	Stiff clay	150-300
	Firm clay	75-150
	Soft clay	75

Soil survey

Purpose

Soil surveys may be required to determine either, or both, of the following:-

- ❖ Environmental contamination.
- ❖ Geotechnical properties.

Environmental

An intrusive investigation may be required to establish the nature of any sub-surface contaminated soil and/or groundwater. Site investigations are based upon site-specific factors such as the presence of likely contaminants, the anticipated sub-surface geology and the location of the site. They are often referred to as Phase 2 investigations where they follow a non-intrusive Phase 1 audit. Samples of soil and groundwater are taken for laboratory analysis. The results are interpreted and a risk assessment undertaken for any contamination identified.

Geotechnical

The purpose of a geotechnical site investigation provides sufficient geotechnical information to enable design of foundations. Investigations will be designed specifically to suit the anticipated geology and development proposals. A geotechnical site investigation is intrusive and may include in-situ and laboratory tests.

A geotechnical site investigation is often procured with an environmental site investigation given a significant overlap with the fieldwork requirements.

Architectural, engineering and services design

Fieldwork

The following methods of soil and groundwater sampling are common to both environmental and geotechnical investigations.

Trial pits

Extremely valuable if the depth of investigation is less than about 3 metres. Trial pits allow a detailed examination of the ground conditions in-situ with some indication of stability and groundwater conditions. Trial pitting is a relatively fast and efficient means of exploring sub-surface conditions.

Auger holes

Auger holes are normally made by hand-turning a very light auger into the ground, or by using light power-auger equipment. However, while auger holes are effective in preliminary investigations they do not provide the depth or range of sampling and in-situ testing provided by conventional site investigation boreholes.

Window samplers

A window sampler is a steel tube, usually about 1 metre long, with a series of windows along the tube through which to view disturbed soil conditions or extract samples. A lightweight percussion hammer drives the sampler into the ground, which is then extracted with jacks. A depth of approximately 10 metres can be achieved, depending on soil conditions, using a sequence of progressively smaller diameter samplers.

Boreholes

Light percussion drilling (shell and auger) is the most commonly used method in the UK. Samples are often unsuitable for soil description on the basis that the drilling process will have changed the strength of the soil by remoulding it and increased the moisture content of the soil through lubrication. For this reason U100 tube samples are often taken from boreholes at regular depths by driving a small diameter tube into the soil at the base of the borehole, thus extracting an undisturbed sample for laboratory analysis. Lubrication may be required if the sub-surface is not cohesive making it unsuitable for some types of environmental site investigations.

Geotechnical in-situ testing

In-situ testing is typically undertaken on cohesionless soils where the strength characteristics may only be established on undisturbed in-situ samples.

Cone Penetration Test (CPT)

This technique involves hydraulically pushing a 10 or 15cm^2 cone into the ground at a standard rate of penetration and measuring the penetration resistance. The equipment necessary for this type of investigation is housed in a large truck. The results from a CPT can be very valuable providing a soil profile, estimates of the soil types encountered including consistency and density, and evidence of the presence of voids beneath the site.

Dynamic probing

Much less sophisticated than CPT, dynamic probing involves driving a steel rod into the ground by using repeated blows of a hammer of a specified mass falling through a fixed distance. The number of blows required for each 100mm is recorded and plotted as a depth versus blow-count log. However, the information given by dynamic probing is very restricted and is difficult to interpret. Further investigations are usually necessary to supplement this test.

The Standard Penetration Test (SPT)

This test involves driving an open-drive sampler into the bottom of a borehole with repeated blows of a hammer falling a predetermined distance. The number of blows necessary to drive the sampler six increments of 75mm are counted and recorded, giving the penetration resistance value. The test is undertaken on cohesionless soils such as gravels.

Other methods of testing and sampling are used depending on the soil type and conditions on site.

Geotechnical laboratory testing

Laboratory testing is typically undertaken on cohesive soils ie clays, where an undisturbed sample may be retrieved from the field. Such testing is not appropriate to cohesionless soils such as gravels. Tests include:-

- ❖ Soil Classification: Particle Size Distribution, Plasticity Index, Moisture Content.
- ❖ Consolidation: To assess the settlement characteristics.
- ❖ Triaxial: To assess the compressive strength characteristics.

Geotechnical reporting

The results of the fieldwork and laboratory tests are assessed. For development purposes preliminary designs for foundations may be prepared. Sufficient information would be provided to enable design of building foundations. For diagnostic purposes the investigation would assist with identifying the cause of defects with a building superstructure allowing remedial solutions to be specified with confidence.

Environmental sampling and analysis

Soil sampling

Soil samples are required for chemical analysis to determine the presence of any contamination. Methods of soil sampling differ depending on the contaminants being tested. Usually disturbed samples are adequate for testing most chemicals. These can be obtained from the excavator bucket when trial pits are being used, from the cuttings from boreholes or from the window within a window sampler.

Samples are analysed in the laboratory for a wide range of 'base line' contaminant chemicals often supplemented by further specialist testing depending upon the type of contamination present.

Water sampling

Water samples can be obtained by inserting a standpipe into a completed borehole. Water samples will be taken after insertion for further chemical analysis. The well may also be used for the monitoring of gas. Installation can be permanent/semi-permanent to facilitate further sampling at a later date. Laboratory analysis is similar to soil samples.

Gas sampling

The presence of methane and carbon-dioxide may be established on sites which are landfills or close to neighbouring landfills. Monitoring wells constructed for water sampling may also be used to sample the presence of land gas. In-situ measurements of gas concentrations and flow are taken at ground level on the head of the monitoring well using an infra-red gas analyser. Wells may be constructed to measure gas concentrations and flow at different depths.

A simpler broad-brush method of establishing the general presence of land gas is to undertake a shallow 'spike survey'. The presence and extent of contamination from a leaking underground petrol tank is a particularly common example.

Architectural, engineering and services design

Environmental risk assessment

On completion of the fieldwork and laboratory testing a risk assessment is undertaken. Where contamination has been identified it is important to consider:-

- The client's requirements.
- The present use of the site and risk to occupants.
- The future intended use of the site and any considerations triggered by redevelopment.
- Future excavations for foundations or buried services.
- Contamination of groundwater particularly where abstracted for water supply.
- Contamination of nearby water courses.
- Contamination to or from neighbouring sites.

Recommendations and budget costs should be provided for remediation.

Design loadings for buildings

The following tables (Tables 1 and 4 from BS6399: Part 1) indicate minimum recommended imposed loads for use in the design of certain parts or types of building. For the expanded list and limitations of use, reference should be made to BS 6399: Parts 1 and 3: 1996.

Architectural, engineering and services design

Table 1. Minimum imposed floor loads

Type of activity/ occupancy part of the building or structure	Examples of specific use		Uniformly distributed load kN/m²	Concentrated load kN
A Domestic and residential activities (Also see catagory C)	All usages within self-contained dwelling units communal areas (including kitchens) in blocks of flats with limited use (See note 1) (For communal areas in other blocks of flats, see C3 and below)		1.5	1.4
	Bedrooms and dormitories except those in hotels and motels		1.5	1.8
	Bedrooms in hotels and motels Hospital wards Toilet areas		2.0	1.8
	Billiard rooms		2.0	1.8
	Communal kitchens except flats covered by note 1		3.0	4.5
	Balconies	Single dwelling units and communal areas in blocks of flats with limited use (See note 1)	1.5	1.4
		Guest houses residential clubs and communal areas in blocks of flats except as covered by note 1	Same as rooms to which they give access but with a minimum of 3.0	1.5/m run concentrated at the outer edge
		Hotels and motels	Same as rooms to which they give access but with a minimum of 4.0	1.5/m run concentrated at the outer edge
B Offices and work areas not covered elsewhere	Operating theatres, X-ray rooms, utility rooms		2.0	4.5
	Work rooms (light industrial) without storage		2.5	1.8
	Offices for general use		2.5	2.7
	Banking halls		3.0	2.7
	Kitchens, laundries, laboratories		3.0	4.5
	Rooms with mainframe computers or similar equipment		3.5	4.5
	Machinery halls, circulation spaces therein		4.0	4.5
	Projection rooms		5.0	To be determined for specific use

Architectural, engineering and services design

Table 1. Minimum imposed floor loads *(continued)*

	Examples of specific use	Uniformly distributed load kN/m²	Concentrated load kN
	Factories, workshops and similar buildings (general industrial)	5.0	4.5
	Foundries	20.0	To be determined for specific use
	Catwalks	-	1.0 at 1m centres
	Balconies	Same as rooms to which they give access but with a minimum of 4.0	1.5/m run concentrated at the outer edge
	Fly galleries	4.5 kN/m run distributed uniformly over width	-
	Ladders	-	1.5 rung load
Type of activity /occupancy part of the building or structure	**Examples of specific use**	**Uniformity distributed load kN/m²**	**Concentrated load kN**
C Areas where people may congregate	Public, institutional and communal dining rooms and lounges, cafes and restaurants (See note 2)	2.0	2.7
C1 Areas with tables	Reading rooms with no book storage	2.5	4.5
	Classrooms	3.0	2.7
C2 Areas with fixed seats	Assembly areas with fixed seating (See note 3)	4.0	3.6
	Places of worship	3.0	2.7
C3 Areas without obstacles for moving people	Corridors, hallways, aisles, stairs, landings etc. in institutional type buildings (not subject to crowds or wheeled vehicles), hostels, guest houses, residential clubs and communal areas in blocks of flats not covered by note 1. (For communal areas in blocks of flats covered by note 1, see A)	Corridors, hallways, aisles etc. (foot traffic only) 3.0 Stairs and landings (foot traffic only) 3.0	4.5 4.5
	Corridors, hallways, aisles, stairs, landings etc. in all other buildings	Corridors, hallways, aisles etc (foot traffic only) 4.0	4.5
	hotels and motels and institutional buildings	Corridors, hallways, aisles etc, subject to wheeled vehicles, trolleys etc. 5.0 Stairs and landings (foot traffic only) 4.0	4.5 4.0

Architectural, engineering and services design

Table 1. Minimum imposed floor loads (continued)

	Industrial walkways (light duty)	3.0	4.5
	Industrial walkways (general duty)	5.0	4.5
	Industrial walkways (heavy duty)	7.5	4.5
	Museum floors and art galleries for exhibition purposes	4.0	4.5
	Balconies (except as specified in A)	Same as rooms to which they give access but with a minimum of 4.0	1.5/m run concentrated at the outer edge
	Fly galleries	4.5 kN/m run distributed uniformly over width	-
C4 Areas with possible physical activities (See clause 9)	Dance halls and studios, gymnasia, stages	5.0	3.6
	Drill halls and drill rooms	5.0	9.0
C5 Areas susceptible to overcrowding (See clause 9)	Assembly areas without fixed seating, concert halls, bars, places of worship and grandstands	5.0	3.6
	Stages in public assembly areas	7.5	4.5
D Shopping areas	Shop floors for the sale and display of merchandise	4.0	3.6
Type of activity/ occupancy part of the building or structure	Examples of specific use	Uniformity distributed load kN/m²	Concentrated load kN
E Warehousing and storage areas. Areas subject to accumulation of goods. Areas for equipment and plant	General areas for static equipment not specified elsewhere (institutional and public buildings)	2.0	1.8
	Reading rooms with book storage, eg, libraries	4.0	4.5
	General storage other than those specified	2.4 for each metre of storage height	7.0
	File rooms, filing and storage space (offices)	5.0	4.5
	Stack rooms (books)	2.4 for each metre in storage height but with a minimum of 6.5	7.0
	Paper storage for printing plants and stationery stores	4.0 for each metre of storage height	9.0

111

Architectural, engineering and services design

Table 1. Minimum imposed floor loads (continued)

	Dense mobile stacking (books) on mobile trolleys, in public and institutional buildings	4.8 for each metre of storage height but with a minimum of 9.6	7.0
	Dense mobile stacking (books) on mobile trucks, in warehouses	4.8 for each metre of storage height but with a minimum of 15.0	7.0
	Cold storage	5.0 for each metre of storage height but with a minimum of 15.0	9.0
	Plant rooms, boiler rooms, fan rooms, etc, including weight of machinery	7.5	4.5
	Ladders	-	1.5 rung load
F	Parking for cars, light vans, etc. not exceeding 2500 kg gross mass, including garages, driveways and ramps	2.5	9.0
G	Vehicles exceeding 2500 kg. Driveways, ramps, repair workshops, footpaths with vehicle access, and car parking	To be determined for specific use	

NOTE 1. Communal areas in blocks of flats with limited use refers to blocks of flats not more than three storeys in height and with not more than four self-contained dwelling units per floor accessible from one staircase.

NOTE 2. Where these same areas may be subjected to loads due to physical activities or overcrowding, eg, a hotel dining room used as a dance floor, imposed loads should be based on occupancy C4 or C5 as appropriate. Reference should also be made to clause 9.

NOTE 3. Fixed seating is seating where its removal and the use of the space for other purposes is improbable.

Architectural, engineering and services design

Table 4. Minimum horizontal imposed loads for parapets barriers and balustrades, etc

Type of occupancy for part of the building or structure	Examples of specific use	Horizontal uniformly distributed line load (kN/m)	A uniformly distributed load applied infill (kN/m^2)
A Domestic and residential activities	(i) All areas within or serving exclusively one dwelling including stairs, landings, etc. but excluding external balconies and edges of roofs (see C3 ix)	0.36	0.5
	(ii) Other residential, (but also see C)	0.74	1.0
B and E Offices and work areas not included elsewhere including storage areas	(iii) Light access stairs and gangways not more than 600mm wide	0.22	N/A
	(iv) Light pedestrian traffic routes in industrial and storage buildings except designated escape routes	0.36	0.5
	(v) Areas not susceptible to overcrowding in office and institutional buildings also industrial and storage buildings except as given above	0.74	1.0
C Areas where people may congregate	(vi) Areas having fixed seating within 530mm of the barrier, balustrade or parapet	1.5	1.5
C1/C2 Areas with tables or fixed seating	(vii) Restaurants and bars	1.5	1.5
C3 Areas without obstacles for moving people and not susceptible to overcrowding	(viii) Stairs, landings, corridors, ramps	0.74	1.0
	(ix) External balconies and edges of roofs. Footways and pavements within building curtilage adjacent to basement/sunken areas	0.74	1.0
C5 Areas susceptible to overcrowding	(x) Footways or pavements less than 3 m wide adjacent to sunken areas	1.5	1.5
	(xi) Theatres, cinemas, discotheques, bars, auditoria, shopping malls, assembly areas, studio. Footways or pavements greater than 3 m wide adjacent to sunken areas	3.0	1.5
	(xii) Designated stadia (see note 1)	See requirements of the appropriate certifying authority	
D Retail areas	(xiii) All retail areas including public areas of banks/building societies or betting shops. For areas where overcrowding may occur, see C5	1.5	1.5

Architectural, engineering and services design

Table 4. Minimum horizontal imposed loads for parapets barriers and balustrades, etc *(continued)*

F/G Vehicular	(xiv) Pedestrian areas in car parks including stairs, landings, ramps, edges or internal floors, footways, edges of roofs	1.5	1.5
	(xv) Horizontal loads	See Clause	11

NOTE 1. Designated stadia are those requiring a safety certificate under the Safety of Sports Ground Act 1975

Building services design

- Air conditioning systems
- Plant and equipment
- Lift terminology
- Lighting design
- Data installations

Architectural, engineering and services design

Architectural, engineering and services design

Air conditioning systems

Why air condition?
- ❖ market requirements
- ❖ location influence
- ❖ environmental restraints
- ❖ tenant requirements
- ❖ internal heat gains

Systems

Direct expansion (DX) packaged systems
- ❖ unitary or split type arrangements possible
- ❖ split systems can serve a limited number of rooms
- ❖ low capital cost with medium running costs
- ❖ poor flexibility for change
- ❖ can operate in reverse cycle heat pump mode
- ❖ reasonable environmental control
- ❖ average resultant room noise levels
- ❖ limited riser requirements
- ❖ separate space required for outdoor condensers
- ❖ medium maintenance requirements

General use: retail units, restaurants, single offices (ie meeting rooms).

Variable refrigerant volume (VRV) systems
- ❖ perimeter or ceiling mounted
- ❖ up to 32 indoor room units per system
- ❖ normally provided with separate fresh air system
- ❖ individual or group control of room units available
- ❖ medium capital cost with low/medium running costs
- ❖ average flexibility for change
- ❖ reasonable environmental control
- ❖ average resultant room noise levels
- ❖ small service riser plus fresh air requirements
- ❖ separate space required for outdoor units and fresh air plant
- ❖ medium maintenance requirements

General use: small/medium hotels, small/medium office developments, refurbishments with limited floor/ceiling space.

Unitary reverse cycle heat pump systems
- ❖ usually perimeter mounted
- ❖ normally provided with separate fresh air system
- ❖ medium capital cost with medium running costs
- ❖ reasonable/good flexibility for change
- ❖ reasonable environmental control
- ❖ higher resultant room noise levels

Architectural, engineering and services design

- ❖ small service riser plus fresh air requirements
- ❖ separate space required for heating, heat rejection and fresh air plant
- ❖ high maintenance requirements

General use: small/medium office developments.

Four pipe fan coil unit systems
- ❖ perimeter or ceiling mounted
- ❖ normally provided with separate fresh air system
- ❖ medium capital cost with medium/high running costs
- ❖ good flexibility for change
- ❖ good environmental control
- ❖ average resultant room noise levels
- ❖ small/medium service riser plus fresh air requirements
- ❖ separate space required for heating, refrigeration and fresh air plant
- ❖ medium/high maintenance requirements

General use: hotels, medium/large office developments, Institutional speculative developments.

Variable air volume (VAV) systems
- ❖ normally ceiling mounted
- ❖ fresh air integral
- ❖ heating can be integrated or separate at perimeter
- ❖ free cooling available during low ambient conditions
- ❖ medium/high capital cost with low/medium running costs
- ❖ very good flexibility for change
- ❖ good environmental control
- ❖ good resultant room noise levels
- ❖ large service riser requirements
- ❖ separate space required for heating, refrigeration and central air plant
- ❖ medium/high maintenance requirements

General use: medium/large size principal office developments.

Underfloor all-air systems
- ❖ more suited to open plan areas
- ❖ fresh air integral with good air quality
- ❖ heating normally required at perimeter
- ❖ free cooling available during low ambient conditions
- ❖ requires good floor void construction
- ❖ low/medium capital cost with low/medium running costs
- ❖ very good flexibility for change
- ❖ good environmental control
- ❖ very good resultant room noise levels
- ❖ medium service riser requirements
- ❖ separate space required for heating, refrigeration and central air plant

Architectural, engineering and services design

- low maintenance requirements

General use: medium/large size office developments, call centres.

Displacement ventilation systems

- more suited to open plan areas with high ceilings
- air inlet at low level perimeter or within raised floor
- fresh air integral with good air quality
- low cooling capability (unless supplemented with chilled beams/ceilings)
- building envelope needs to limit solar gains
- heating normally required at perimeter
- free cooling available during low ambient conditions
- requires good floor void construction
- low capital cost with low running costs
- very good flexibility for change
- good environmental control
- very good resultant room noise levels
- medium service riser requirements
- separate space required for heating, refrigeration and fresh air plant
- low maintenance requirements

General use: medium/large office developments.

Chilled beam/ceiling systems

- more suited to open plan areas
- requires integration within suspended ceilings
- requires separate mechanical fresh air system
- requires low relative humidity control
- active beams integrated with fresh air system
- heating normally at perimeter or function of active beams
- medium/high capital cost with low/medium running costs
- poor flexibility for change
- very good environmental control
- very good resultant room noise levels
- medium service riser requirements
- separate space required for heating, refrigeration and fresh air plant
- low maintenance requirements

General use: medium/large office developments.

Mixed mode ventilation systems

Mixed mode is a term used to describe engineering strategies that normally combine natural ventilation and/or mechanical ventilation and/or cooling in various combinations to achieve acceptable indoor environmental conditions in the most effective manner. It involves maximising the use of the building fabric and envelope to modify the internal climate/temperature swings, such as the use of night time pre-cooling of the building structure via the mechanical fresh air system. This approach has generally been used in offices; however, it is suitable for a wide range of building types.

Architectural, engineering and services design

Plant and equipment

Boilers – (purpose is to heat water for distribution)
- ❖ fuel can be gas, oil, coal, electric or dual fuel
- ❖ air is required for combustion and cooling (not for electric)
- ❖ some form of flue is required (not for electric)
- ❖ all sizes are readily available
- ❖ usually quiet in operation with insignificant vibration

Water chillers – (purpose is to cool water for distribution)
- ❖ normally electrically driven, but can be gas driven
- ❖ conventionally air cooled to avoid the need for cooling towers that can be susceptible to bacteriological contamination (if near large body of water, water cooled chiller can be used)
- ❖ best externally located but can be internal with large external heat rejection, if essential
- ❖ all sizes are available
- ❖ refrigeration machines are complex and expert maintenance is required
- ❖ high noise and vibration levels can be generated
- ❖ old machines incorporated CFCs and HCFCs but new ones must not

Air handling units – (purpose is to deliver heated/cooled and filtered air to various spaces)
- ❖ electrically driven
- ❖ units incorporate fans, heaters, coolers, humidifiers and filters in various combinations
- ❖ may be located externally or internally with fresh air ducts to outside
- ❖ all sizes available
- ❖ low tech equipment and easily maintained
- ❖ noise and vibration levels can be contained (but additional space required in plant room for silencers)

Diesel generators – (purpose is to provide a standby electrical supply to compensate for a breakdown in the main utility supply)
- ❖ normally diesel oil fed for commercial developments
- ❖ high fresh air and exhaust requirements demanding large louvred areas (for cooling)
- ❖ noise and vibration levels require special attention
- ❖ oil storage facility – daily use and long term storage facility required
- ❖ flue is required

Architectural, engineering and services design

Lift terminology

Power systems

Traction
- ideally requires motor room above (possible planning problem)
- can be positioned at other levels, usually below or adjacent to lift pit – more expensive (doubles load on structure)
- models now available which do not require a motor room
- incorporates counterbalance weight
- high efficiency
- high speed available

Types
- single speed motor: up to 0.5m/sec jolt stop
- dual speed motor: up to 1.0m/sec, more accurate levelling
- geared variable voltage: smoother ride and greater speed
- gearless variable voltages above 2.0m/sec, very quiet, grouped application, long travel

Hydraulic
- higher starting current, but not significant as compared to electric traction lifts
- maximum travel 20m
- less efficient, higher energy consumption
- no counterbalance
- limited starts per hour
- motor and pump house can be remote from shaft (up to 10m from lift pit). (Ventilation important for cooling)
- savings on builder's work
- lower speed, up to 1.0m/sec

Types
- direct acting: ram and bore hole below car
- side acting: Usually 'fork lift' action ram and cylinder within shaft
- indirect: combined ram and ropes, ram raises and lowers pulley

Control systems

Button operation
- operator controlled

Largely superseded

Automatic push button
- responds to first push – no calls stored; preference given to car button calls – flats and small offices

Down collective
- answers landing calls in down direction only – mainly used for flats

Full collective
- calls stored and answered in sequence in both up and down direction

Architectural, engineering and services design

Full collective interconnected
- group of lifts interconnected as in full collective

Programmed (or group collective) – tending to supersede other types.

Selector
Monitors position of lift car in shaft. Type varies with manufacturer.
Pulse selector – using proximity switches.

Mechanical selector
- cable linked to 'model' in control panel
- cable linked to carrier
- shaft switches

Door arrangement

Manually operated

Power operated
- single sliding: cheapest – larger shaft size required
- two speed sliding: expensive – minimum shaft size
- centre opening: quickest to full opening position

Safety provisions
- door safety edge – operating micro-switch on door
- pressure sensitive doors – operating micro switch on door operating mechanism
- electronic proximity detectors – detects presence of obstruction
- light ray or sonar

For further information refer to BS EN81 Parts 1 and 2.

Lighting design

Category	Lux	Typical areas
Casual	100-150	Storage areas, plant rooms, lifts, circulation areas, bathrooms
Casual rough work	200-300	Dining areas, lounging rooms, bars, sports halls, libraries, rough machining
Routine work	300-500	General office, retail areas, lecture rooms, laboratories, kitchens, medium machining, supermarkets
Demanding work	750	Drawing offices, inspection of medium machining
Detailed work	1,000	Colour discrimination, fine machining and assembly, inspection rooms
Very fine work	1,500-3,000	Hand engraving, precision works, inspection of fine works

Architectural, engineering and services design

Lighting in areas where display screens are in use should comply with the requirements of CIBSE Lighting Guide LG3, including 'Addendum 2001' released in October 2001. The addendum covers new luminance limits for the performance of luminaires relating to various display screen types, in place of the old Cat 1, 2 and 3 standards which are now withdrawn. The purpose of the addendum is to promote consideration being given to the overall visual environment (ie surface reflections, direct daylight etc) and its effect on display screens and their users. Emphasis is also placed on designing schemes that avoid very high luminance patches in a space and abrupt changes in luminance across a surface or between adjacent surfaces.

A certificate of conformity to LG3 was introduced from May 2002 for completion by both the designers and installers. Copies of the certificate can be obtained from the CIBSE website.

Data installations

Data installations and information technology require extensive cabling which, in turn, demands adequate access through the building. Therefore, it is important to consider access via risers, suspended floors or floor trunking in order to present the user with these services.

A raised floor is normally a basic pre-requisite to enable full flexibility of outlet positions. Minimum clear void typically 100 – 150mm depending upon floor plate size and riser arrangement.

Modern buildings are now usually block wired for voice and data using common telephone/data jack outlets.

Data/IT cabling is likely to require frequent upgrading and therefore components and component wiring needs to be accessible.

Site analysis

Measured surveys

Architectural, engineering and services design

Architectural, engineering and services design

Measured surveys

It is strongly advisable to invest in measured surveys at all stages of a project, as early collection of the correct data will pay dividends.

The surveys are best carried out by a reputable measurement surveyor with independence and integrity. Most projects can be broken down into the following stages – the type of surveys recommended at each stage are listed.

Property acquisition – Check the floor areas independently (the RICS produce a booklet 'Code of Measurement' which is worth reading).

Check the lease or conveyance plans with the latest Ordnance Survey (OS) maps (1:1250 or 1:2500 scale). Check boundaries on site against legal documentation – resolve boundary disputes before completing the contract.

Development – Existing measured surveys of the topography and/or buildings may exist. If they do, be sure to check their completeness and whether they are up to date. Consider employing a land surveyor to make checks. If none exist, then purchase the latest OS mapping (available digitally) for contextual purposes but be aware that the data will only be accurate to a metre or two and may not show all relevant features due to the scale.

It is strongly advised that you invest in a new measured survey. The RICS or TSA can provide assistance. The RICS has a document for large scale surveys which needs you to 'tick the boxes' to provide a specification - it is not a specification without you doing this. Also be aware that the more boxes you tick - the more cost is involved.

Various techniques are available now for obtaining data for adjoining properties such as large scale aerial photography and laser scanning (LIDDAR). This enables 3D data to be collected without gaining access to the property. Various 3D models of cities are now available on the internet but be very wary of their accuracy.

Architectural engineering and services, design and construction

The architect, engineer and services consultant will need measurements.

Everyone can and does take measurements. Modern equipment, such as Leica's 'Disto' and electronic theodolites make the task of surveying ever easier. It is strongly advisable though to seek the services of a good, reputable and honest land surveyor whose job it is to 'measure'. He will understand the complexities of the instrumentation and the necessary precision and accuracy for the task.

The culture within the property and construction industry is not to rely on others' dimensions and measurements. This all too often leads to disputes over measurements. It is worth considering employing an independent surveyor to collect the necessary measurements and supply these to all parties involved. A good surveyor will stand by his dimensions and this will lead to an improvement in time and less confrontation which should inevitably lead to cost savings. The new laser scanning technology provides an accurate 3D model instantly, from which measurements can be taken. This can be used to record progress and be displayed at design team meetings for all to see and take measurement from. Due to the fact this is 3D there is instant 'clash detection' (important for M&E Services), there is less ambiguity and misreading than with 2D drawings and finally (and most importantly) it is understood by non-technical professions (eg, finance and legal advisers).

During construction it will be advisable to have a land surveyor on site, all the time if possible, or at least at regular intervals to ensure adequate

Architectural, engineering and services design

dimensional control is established (in X, Y and Z) for the trade contractors to use.

After completion of the building a true 'as-built' survey is recommended. This can be used to check the area (gross and net) of the building and then be retained as part of the log book of the property. Ensure this is available in digital form to be used in CAD systems as well as a format suitable for Microsoft Office.

The plans will help the maintenance of the building and any future disposal. It is surprising how few reliable drawings of relatively new buildings are found by staff working on the property.

An information system (like GIS) is worth considering to help the operation of the property. Each space can be allocated a unique reference number attached to the drawings to enable instant reporting and, for example, dependent sizes and location. Additional data can be added about size, condition, occupancy, etc.

Computer aided design

Computer aided design

Architectural, engineering and services design

Architectural, engineering and services design

Computer aided design

Computer Aided Design (CAD) has become a widely used tool within the construction industry.

Both construction professionals and their clients now use CAD in the design, procurement and development of a variety of projects. It is this versatility, coupled with ever-reducing costs of hardware and software, that have increased the popularity of CAD and contributed to the decline of the traditional drawing board.

Hardware

The computing power required to run the most up-to-date CAD programs tends to vary between each package. The following system specification is for a typical basic installation, but advice should always be sought prior to purchase:

- Intel® Pentium® II or AMD K611 with a minimum processor speed of 450MHz.
- Microsoft® Windows® XP Professional, Microsoft® Windows® 2000 Professional, Windows 98, Windows ME or Windows NT® 4.0 (SP5 or later). (This software operates on a PC platform, however, many CAD packages have also been developed for use on a MAC platform).
- 128MB RAM and at least 200MB of free disk space (To install and run the software, further disk space will be required to save the drawing and backup files, either on the PC's hard drive or on a file server connected to a Wide or Local Area Network (WAN or LAN).
- VGA display 1024 x 768 or higher.
- CD ROM drive.
- Mouse or other pointing device (some CAD packages can be used with a tablet which provides the user with each command, symbol and block in view on the desktop for easy selection).

Higher specification PCs are available that will enable the user to operate the system more effectively and reduce the time spent to produce drawings.

Software

Since the initial development of desktop systems, a number of software manufacturers have established a market share in the supply of CAD programs, with the most commonly used being AutoCAD®, Microstation and ArchiCAD. Other programs exist which enable the user to produce enhanced drawings, such as 3D rendered visualisations and mechanical and electrical services drawings. Basic packages such as Volo™ View Express are available to download freely from the internet and enable PC users to view, mark up and print digital drawings received without having to purchase and understand a blown CAD program.

Further specialist software packages, which bolt onto the main CAD package, are available to support specialist uses. Examples include right to light calculation, daylight and sunlight analysis, sun path diagrams, digital measured and condition surveys, 3D development images, animations and fly-throughs.

Architectural, engineering and services design

Principles of CAD

The production of two-dimensional drawings with the aid of CAD is based on the age old tradition of drafting on drawing board. The user still requires a good understanding of the principles behind producing a drawing, although the tools and processes used distinctly vary. Functions and terminology common to most CAD programs are listed below.

CAD files come in a variety of formats, such as DXF, DWG, DGN and DWF. When requesting digital drawings from an outside source, it is essential that a compatible format is specified. Drawings produced on a later version of the same software may not always be accessible through an earlier version.

Scale
: Drawings are often prepared at 1:1 scale and then viewed on the title sheet within 'viewports' at a more appropriate scale, such as 1:50. The ability to produce a building or its components in this way increases the level of accuracy at which the drawing is produced and enables details to be generated from the 'master' drawing without having to redraw at an alternative scale.

Layers
: Layers are used within CAD to enable different elements of a building's structure to be separated from another. For example, walls are placed on a different layer to windows. Layers can be turned on or off, removed from the printed copy, assigned a different colour, line thickness or style, without the user having to select each individual element or delete items from the drawing.

Styles
: The styles in which a drawing is produced can be customised on a number of levels. They can be customised throughout a company, for an individual client, an individual project or down to the preferences of each user. This can include the style, the content and setup of each drawing, even the characterisation of the printed drawing copies, ensuring that each drawing issue maintains a uniformity in line with the degree of customisation implemented.

Templates
: Templates enable the user to standardise drawing production. Many building product manufacturers now provide CAD-compatible 'blocks', which can be inserted into a drawing. Subscription services are also available, such as RibaCAD, which produce regular distributions incorporating the latest company templates and blocks.

Raster images
: Raster images, can be incorporated into the drawing to present information which could not otherwise be drawn, for example, photographs.

Plotting
: CAD drawings can be plotted at any scale and on any paper size. Standard paper sizes typically range from between A4 and A0, and scales range from 1:1 for intricate details through to 1:2500 for location plans. Drawings can be plotted in full colour, black and white and/or greyscales. Different line thicknesses can be used to highlight different elements of the drawing.

This list is not exhaustive and only provides a general overview of the simpler functions available in CAD.

The use of CAD in a project

CAD is usually implemented at project inception phase. This may be in the form of a digital survey of the development site and its surroundings. The project architect can then produce initial appraisal and feasibility studies prior to commencement of the full scheme.

Architectural, engineering and services design

As the project develops, CAD can assist the entire project team in the production of project documentation. Compatibility of systems is required to ensure continuity across the team and to facilitate document exchange by e mail, project extranets and web-based file interchanges.

The ability to produce 3D images and models greatly improves presentation quality. Unlike a physical model, a 3D CAD model can be altered and manipulated with ease.

Throughout the construction phase, further working details can be produced and the construction drawings altered readily. Upon completion of the project, 'as-built' drawings can easily be produced and a full set of revisions can be stored.

Advantages and disadvantages of CAD

Advantages

- Improved accuracy
- Greater degree of detail
- Greater complexity
- Multiplication of information
- Speed and efficiency of drawing production
- Ease and speed of drawing distribution
- 3D visualisations, conceptual designs and virtual reality presentations.
- Design team work from the same benchmark
- Design data management
- Paperless project procurement and information storage

Disadvantages

- Degree of accuracy can make sketches long-winded and inefficient
- Drawing production requires trained CAD technicians
- Effective document exchange among project team requires a degree of expertise in IT and appropriate systems in each office
- Incompatibility between drawings produced on CAD packages from different software manufacturers

Further information and hints

www.cadline.co.uk provides details of a large proportion of the CAD software currently available, including the sale and retail of software, training courses available and technical support for a number of CAD packages.

What to look for in CAD software

- ❖ The software package should have a significant share in the market.
- ❖ The software's functions should be sufficient to meet your needs.
- ❖ Ease of learning, and availability of the technical support and training courses.
- ❖ Price.
- ❖ Operating system requirements, including any bolt-on packages required and future upgrades.
- ❖ Ability to work with the data produced in other programs and its compatibility with data provided by others.
- ❖ The software's conformity with industry data standards and reference sites.
- ❖ The long term development of the software and the manufacturers commitment to upgrade the software in line with hardware.

Tables and statistics

Conversion formulae

Architectural, engineering and services design

Conversion formulae

To convert to metric, multiply by the factor shown
To convert from metric divide by the factor shown

Length

miles: kilometres	1.6093
yards: metres	0.9144
feet: metres	0.3048
inches: millimetres	25.4

Area

square miles: square kilometres	2.59
square miles: hectares	258.999
acres: square metres	4046.86
acres: hectares	0.4047
square yards: square metres	0.8361
square feet: square metres	0.0929
square inches: square millimetres	645.16

Volume

cubic yards: cubic metres	0.7646
cubic feet: cubic metres	0.0283
cubic inches: cubic centimetres	16.3871

Capacity

gallons: litres	4.546
US gallons: litres	3.785
quarts: litres	1.137
pints: litres 0.568 gills: litres	0.142

Mass

tons: kilogrammes	1016.05
tons: tonnes	1.0160
hundredweights: kilogrammes	50.8023
quarters: kilogrammes	12.7006
stones: kilogrammes	6.3503
pounds: kilogrammes	0.4536
ounces: grammes	28.3495

To convert °C to °F

°C = 5/9 (°F) − 32
°F = 9/5 (°C) + 32

Materials

- Deleterious materials
- Asbestos
- High Alumina Cement concrete
- Chlorides
- Corrosion of metals

Materials and defects

Deleterious materials

The presence of deleterious materials in a building may affect its market value and could, in severe cases, result in element failure or affect the health of persons working or living there.

The reaction of investing institutions to these materials depends on a number of factors and often the presence of a deleterious substance will not prevent a purchase. However, great care must be taken to assess the actual risks or consequences involved, so that a value judgement can be made.

Materials hazardous to health

Materials	Common Use	Use Risk
Lead	When used in water pipes and lead paint (lead roofing materials pose little or no risk)	Risk of contamination of drinking water in lead pipes, or from lead solder used in plumbing joints. Risk of inhalation of lead dust during maintenance of lead based paint. Risk to children of chewing lead painted surfaces. Concentration of lead in paint now generally much reduced
Urea Formaldehyde foam	Cavity wall insulation. Some insulation boards but rare in UK	There is some evidence that UF foam may be a carcinogenic material although this is not proven. Vapour can cause irritation. Poorly installed insulation can lead to passage of water from outer leaf of brick to inner leaf in cavity wall situation. There are some worries over formaldehyde used as an adhesive in medium density fibreboard and chipboard but this is likely to be a problem only in unventilated areas with large amounts of boarding
Asbestos Further details on page 136	Commercial and residential buildings as boarding, sheet cladding, insulation and other uses particularly in the 1950s, 60s and 70s	Airborne asbestos fibres may be inhaled and eventually lead to either asbestosis, lung cancer or mesothemelioma

Materials which may affect building performance or structure

Materials	Common Use	Use Risk
Calcium silicate brickwork	Used in lieu of concrete or clay bricks, often as an inner leaf in cavity work. Often cited as deleterious but if used correctly will perform well	Brickwork shrinks after construction with further movement due to wetting. Construction must provide measures of control to distribute cracking. Concrete bricks may display a similar propensity to shrinkage and again care must be taken in the design of movement joints etc
Calcium chloride concrete additive	Commonly used in in-situ concrete as an accelerator and often added in flake form. Often found in buildings constructed before 1977. (May also be present from atmospheric or traffic exposure)	Reduces passivity of concrete in damp conditions. Subsequent risk of corrosion of steel re-inforcement
High Alumina Cement (HAC) Further details on page 141	Mainly used in the manufacture of pre-cast X or I roof or floor beams together with some lintels, sill members etc between 1954 and 1974. HAC has existed since about 1925	Strength of concrete can decrease significantly often when high temperatures and/or high humidity is involved. Defects may be due to faulty manufacture
Sea dredged aggregate	In-situ concrete or pre-cast concrete	May contain salts such as sodium chloride. If salts not properly washed out, risk of corrosion reinforcement sodium may contribute to alkali silica reaction. Provided the aggregates are properly washed and controlled in accordance with British Standard requirements the indications are that there are no greater risks involved than with the use of aggregates from inland sources

Materials and defects

Materials	Common Use	Use Risk
Mundic blocks and Mundic concrete	Concrete blocks and concrete manufactured from quarry shale commonly found in the West Country	Loss of integrity in damp conditions. Further research required to identify level of risk across the country
Woodwool slabs (also woodcrete and chipcrete)	Often used as (a) decking to flat roofs, or (b) as permanent shuttering	Use in (a) may be considered acceptable providing material is kept dry. Use in (b) as a permanent shutter may result in grout loss (honeycombing) or voidage of concrete near to or surrounding reinforcement, particularly with ribbed floors. May result in reduced fire resistance, reinforcement corrosion or in extreme cases loss of structural strength. May be repaired by application of sprayed concrete. Condition investigated by cut-out removal of woodwool in many locations
Brick slips	Typically 1970s and 1980s to conceal flow nibs in cavity walls	Risk of poor adhesion, lack of soft joints can transfer load to slips and cause delamination

Asbestos

Asbestos is the generic term for several mineral silicates occurring naturally in fibrous form. Because of its various useful properties (resistance to heat, acids and alkalis, and good thermal, electrical or acoustic insulator) it has been extensively used in the construction industry.

Three main types in the UK are Chrysotile (white), Amosite (brown) and Crocidolite (blue). It is used in a variety of forms varying from boards or corrugated sheets to loose coatings or laggings and, generally, the more friable the material, the greater the asbestos content.

Health risk

Inhalation of its microscopic fibres can constitute a serious health risk and is associated with several terminal diseases.

Not all asbestos, irrespective of its circumstances, constitutes an immediate risk, although the effects of possible future disturbance or deterioration must be considered. Factors to be taken into account are the type, form, friability, condition and location of the source material.

The (1987) Joint Central and Local Government Working Party on Asbestos concluded that "Asbestos materials which are in good condition and not releasing dust should not be disturbed... Materials that are damaged, deteriorating, releasing dust or which are likely to do so should be sealed, enclosed or removed as appropriate. Materials which are left in place should be managed and their condition periodically reassessed. The risk to the health of the public from asbestos materials which are in sound condition and which are undisturbed is very low indeed. Substitute

Materials and defects

materials should be used where possible, provided they perform adequately".

The HSE actively discourages the unnecessary removal of sound asbestos materials and each case should be decided on its own merits following an assessment of the risks arising.

In the past the people most at risk have been workers in the asbestos industry involved in the importation, storage, manufacture and installation of materials or components containing this material.

These activities are now banned in the UK and the removal, treatment or intentional working with asbestos is strictly controlled and generally limited to specialists. The risk however continues for anyone who inadvertently disturbs the asbestos in the course of their routine business, including builders, maintenance workers, electricians and the like, and new legislation is intended for their protection.

Legislation

The enabling act for asbestos legislation is the 1974 Health & Safety at Work Act and failure to comply with the requirements is a criminal offence.

There are two principal regulations that apply to works that could expose persons to the risk of respirable asbestos fibres.

The Asbestos (Licensing) Regulations 1983 (as amended)

Broadly, these regulations require any person who manages or carries out work with asbestos insulation, asbestos coating or asbestos insulating board, or the clearance of asbestos contaminated land, to hold a licence that is issued by the HSE.

The Control of Asbestos at Work Regulations (CAWR) 2002

These consolidated similar previous asbestos regulations and they also introduced a new duty to 'manage asbestos' (Regulation 4) together with a requirement to use only accredited persons to analyse the content of materials suspected of containing asbestos (Regulation 20).

With the exception of regulations 4 and 20, CAWR is effective from 21 November 2002. There are transitional periods of 18 months and 24 months respectively for these other two regulations, with regulation 4 coming into force on 21 May 2004 and regulation 20 on 21 November 2004.

Notwithstanding the requirement for a licence, CAWR applies to any work which **may** involve the exposure of persons to **any** form of asbestos irrespective of the type, form or amount of asbestos and includes activities involved in both sampling and laboratory analysis.

Broadly, and using the number of the appropriate regulation as a prefix, they require:-

a) The 'Dutyholder' to:-
4   Manage asbestos in non-domestic premises.

b) 'Every person' to:-
4 (2)   Cooperate with the dutyholder so far as is necessary to enable him/her to comply with their duties to manage asbestos in non-domestic premises.

c) The employer to:-
5   Identify the type of asbestos.
6   Prior to the works, assess the likely level of risk, determine the nature and degree of exposure and set out steps to prevent or control it.
7   Produce a suitable written plan of work.
8   *Notify the Enforcing Authority of non-licensable work (at least 14 days before works).
9   Inform, instruct and train personnel.

Materials and defects

10. Prevent exposure of employees to asbestos so far as is reasonably practicable and where not reasonably practicable, reduce to lowest level reasonably practicable, both the exposure (without relying on the use of respirators) and the number of employees exposed.
11. Ensure that any control measures are properly used or applied.
12. Maintain control measures and equipment (keeping records of the latter).
13. Provide suitable personal protective clothing and ensure that it is properly used and maintained.
14. Establish arrangements to deal with accidents, incidents and emergencies.
15. Prevent spread of asbestos from the workplace, or where not reasonably practicable, reduce to lowest level reasonably practicable.
16. Keep asbestos working areas and plant clean and thoroughly clean on completion.
17. * Designate 'asbestos areas' and 'respirator zones'. Monitor and record exposure where appropriate.
18. *Monitor exposure of employees to asbestos by air monitoring (if not deemed necessary record reasoning).
19. Ensure that air testing is only carried out by a person with ISO 17025 accreditation.
20. Ensure that sampling of material to determine whether it contains asbestos is only carried out by a person with ISO 17025 accreditation.
21. * Maintain health records and medical surveillance of employees.
22. Provide suitable washing and changing facilities (in addition to general welfare).
23. Ensure all raw materials or asbestos waste is stored, received into, dispatched from or distributed within suitable sealed, clearly labelled containers.

   * These specific requirements are only triggered if the likely level of exposure to respirable fibres exceeds stated limits.

The Regulations establish these trigger points in the form of 'action levels' and 'control limits'.

Action levels

If the action level is likely to be exceeded, the following will apply:-

- ❖ Written notification of Enforcing Authority (14 days prior to commencing work).
- ❖ Designation of 'asbestos working areas' where access is limited to authorised personnel.
- ❖ Regular medical surveillance of employees with health records kept for at least 40 years from the date of the last entry.

The action levels refer to the following accumulative exposures whereby the respirable asbestos fibres per millilitre of air are multiplied by the number of hours exposure over a continuous 12-week period.

Type of asbestos	Fibre-hours per ml
White	72
Any other type or mixture	48

Control limits

If the control limit is likely to be exceeded, the following will apply:-

- ❖ Designation of 'respirator zones' where suitable respirators must be worn at all times.
- ❖ Suitable respirators and protective equipment must be issued, used and maintained.

Materials and defects

The control limits are:

Type of asbestos	Average/measured in fibres per ml over a continuous period	
	four hours	ten minutes
White	0.3	0.9
Any other type or mixture	0.2	0.6

"Duty to manage asbestos in non-domestic premises" (regulation 4)

The 'dutyholder' responsible for the management of asbestos in non-domestic premises as set out in regulation 4(1) is every person, who has by virtue of a contract or tenancy, an obligation for its repair or maintenance, or, in the absence of such, control of those premises or access thereto or egress therefrom.

This includes those persons with any responsibility for the maintenance, or control, of the whole or part of the premises.

When there is more than one dutyholder, the relative contribution required from each party in order to comply with the statutory duty, will be shared according to the nature and extent of the repair obligation owed by each.

This regulation does not apply to 'domestic premises', namely a private dwelling in which a person lives, but legal precedents have established that common parts of flats (in housing developments, blocks flats and some conversions) are not part of a private dwelling.

The common parts, are classified as 'non domestic' and therefore regulation 4 applies to them, but not to the individual flats or houses in which they are provided.

Typical examples of common parts are entrance foyers, corridors, lifts, their enclosures and lobbies, staircases, common toilets, boiler rooms, roof spaces, plant rooms, communal services, risers, ducts and external outhouses, canopies, gardens and yards.

The regulation does not however apply to kitchens, bathrooms or other rooms within a private residence, that are shared by more than one household, or communal rooms within sheltered accommodation.

Subject	Requirement
*Cooperate	Cooperate with other dutyholders so far as is necessary to enable them to comply with their Regulation 4 duties
Find and assess condition of ACMs	Ensure that a suitable and sufficient assessment is made as to whether asbestos is or is liable to be present in the premises and its condition, taking full account of building plans or other relevant information, the age of the building and inspecting those parts of the premises which are reasonably accessible
	(Must presume that materials contain asbestos unless strong evidence to the contrary)
	(see MDHS 100 for guidance on asbestos surveys)

Materials and defects

Subject	Requirement
Review	Review assessment if significant change to premises or suspect that it is no longer valid and record conclusions of each review
Records	Keep an up-to-date written record of the location, type (where known), form and condition of ACM's
Risk assessment	Where asbestos is or is liable to be present assess the risk of exposure from known and presumed ACM's
**Management plan	Prepare and implement a written plan, identifying those parts of the premises concerned, specifying measures for managing the risk including adequate measures for properly maintaining asbestos or where necessary, its safe removal
Provide information to others	Ensure the plan includes adequate measures to ensure that information about the location and condition of any asbestos is provided to every person likely to disturb it and is made available to the emergency services
Review and monitor	Regularly review and monitor the plan to ensure it is valid and that the measures specified are implemented and that these are recorded

** Management plan

(see HSE publication 'A comprehensive guide to managing asbestos in premises' HSG227).

The management plan is an important legal document which in addition to its health and safety significance will be required to be made available to, and inspected by, a variety of interested parties.

The absence of such a document may thus have significant financial implications or affect the liquidity of the premises as an asset.

A plan is not required when the assessment whether asbestos is or is liable to be present in the premises confirms that it is not. For example, the building is new and there is confirmation from the project team that asbestos has not been used in its construction.

Nevertheless a record should be kept of the assessment carried out and its conclusion to show to an inspector or prospective purchaser or occupant.

The dutyholder owns and is responsible for the safekeeping of the plan, however he is obliged to make the information available "at a justifiable and reasonable cost" to anyone who is likely to disturb asbestos and this includes new owners or occupants.

Duty to cooperate

'Every person' has a duty to cooperate with the dutyholder so far as is necessary to enable the dutyholder to comply with his duties under Regulation 4.

This includes the landlord, tenants, occupants, managing agent, contractors, designers and planning supervisor.

Materials and defects

The possible scenarios envisaged by the ACOP include:-

- ❖ Anyone with relevant information on the presence (or absence) of asbestos.
- ❖ Anyone who controls parts of the premises to which access will be necessary to facilitate the survey and management of asbestos (ie its removal or treatment or periodic inspection).

Cooperation does not extend to paying the whole or even part of the costs associated with the management of the risks of asbestos by the dutyholder(s), who must meet these personally.

Where there is more than one duty holder for a premise the costs of compliance will be apportioned according to the terms of any lease or contract determining the obligation to any extent for its repair and maintenance.

If there is no such documentation then the apportionment of costs will be based on the extent to which parties exercise physical control over the premises.

In the final analysis, the courts will decide financial responsibility using the principles outlined above.

Guidance in the ACOP states that architects, surveyors or building contractors who were involved in the construction or maintenance of the building and who may have information that is relevant "would be expected to make this available at a justifiable and reasonable cost".

The duty to cooperate is not subject to any limitation or exclusion, thus there is an obligation to do whatever is necessary to cooperate with the dutyholder.

For example, a landlord with a lease covenant that, in the event of the default of the tenant, gives the right to enter and carry out works to ensure compliance with statutory provisions, as a last resort, could be obliged to pursue this option and claim back the costs as part of the service charges.

Short lease tenants, licensees or other occupants who control access, but do not have any contractual maintenance liabilities would be required to permit the landlord access to fulfil his/her duties.

High Alumina Cement concrete

Background

The manufacture of High Alumina Cement (HAC) commenced in the United Kingdom in 1925 to provide concrete that would resist chemical attack, particularly for marine applications. This cement developed high early strength, although its relatively high cost prevented extensive use.

During the late 1950s and 1960s the main use of HAC was for the manufacture of precast prestressed components which could be manufactured quickly, therefore offsetting the additional cost of the material.

The earliest UK failures were experienced during 1973-74 when school roofs collapsed. There followed considerable investigation and testing. In 1975 the Building Regulations Advisory Committee [BRAC] Sub-Committee P published design criteria to be used in checking the adequacy of buildings containing HAC structural members. The two main aspects of the investigation were, firstly, a strength assessment and, secondly, a durability assessment.

In 1976 HAC concrete was banned for structural use.

Materials and defects

Problems with HAC

HAC concrete undergoes a mineralogical change known as conversion. This conversion is accompanied by a loss of strength and increased porosity. Consequently there is also a reduction in resistance to chemical attack. The higher the temperature during the casting of the concrete, the more quickly conversion takes place.

The relationship between conversion and strength is complex, however the strength of highly converted concrete is very variable and is substantially less than its initial strength. Typically the original design strength of 60N/mm2 may be reduced to 21N/mm^2. This lower figure represents a residual strength below which no further loss of strength will occur during the remaining life of the material.

Highly converted HAC concrete is vulnerable to acid, alkaline and sulphate attack. For this to take place water as well as the chemicals must have been present persistently over a long period of time at normal temperatures. Chemical attack is usually very localised in nature and the concrete typically degenerates to a chocolate brown colour and becomes very friable, often due to sulphate attack.

Given the sensitivity to moisture the greatest risk therefore lies in the use of roof members. It is therefore important to appraise the condition of the concrete and any waterproof coverings before making any formal judgement as to the remedial work required.

In a warm and moist environment there is the possibility of chemical action occurring where high alkali levels may be present as a consequence of the use of certain types of aggregate or where alkalis may have ingressed from plasters, screeds and woodwool slabs. In such circumstances HAC is vulnerable to chemical attack.

Investigation

There are three generic stages in an investigation, namely:-

- ❖ Stage 1 - identification
- ❖ Stage 2 - strength assessment
- ❖ Stage 3 - durability assessment

A Stage 2 strength assessment is required to determine if the precast concrete members have sufficient structural capacity, even at the reduced fully converted strength, to safely withstand the applied loading. Sub-Committee P sets out the guidelines for concrete strength based on 21N/mm2. The strength assessment requires the section properties of the beam to be established. Thereafter the structural strength of the element can be calculated. In cases where the section properties are unknown, or cannot be determined by investigation, then assessments are limited to determining the concrete strength using near-to-surface tests.

A Stage 3 durability assessment is required to determine the long-term durability where affected by chemical attack and reinforcement corrosion. Testing can be undertaken to determine the presence of alkalis and sulphates. Laboratory testing may be supplemented by a detailed visual inspection and the removal of lump samples for petrographic examination. A durability assessment should also include a visual examination of the reinforcing steel where lump samples are removed. In recent years it has been found that HAC is less durable then members containing ordinary Portland cement.

Assessment

Putting all of this in context, however, there have been no recorded instances in this country of a failure of a floor incorporating HAC concrete. It is important to know that in the case of the original historic failures, manufacturing faults were eventually discovered. The greatest reduction in strength occurred where a high water content was present during the period of mixing and high temperatures took place during curing.

Of the five failures or new failures of roof constructions, two did not directly involve the quality of the concrete, one was aggravated by chemical attack and two were apparently due to defective concrete which should have been rejected at the time of casting. On no occasion has weakening of the concrete due to conversion been the sole cause of failure.

Chlorides

The presence of chlorides, whether added as calcium chloride or ingressed as de-icing salts, may result in 'chloride induced corrosion' and is less common than corrosion caused by low cover. When chloride corrosion does occur its effects may be wide ranging including a reduction in structural capacity.

Natural alkalinity of concrete

Steel does not corrode when embedded in highly alkaline concrete despite high moisture levels in the concrete because a passive film forms on the steel and remains intact as long as the concrete surrounding the bar remains highly alkaline.

Chloride induced corrosion

Corrosion may occur in concrete that contains sufficient chlorides even if it is not carbonated or showing visible signs of deterioration.

The presence of free chloride ions within the pore structure of the concrete interferes with the passive protective film formed naturally on reinforcing steel.

Chloride ions exist in two forms in concrete namely free chloride ions, mainly found in the capillary pore water, and combined chloride ions which result from the reaction between chloride and the cement hydration process. These occur in proportions that depend on when the chloride entered the concrete. If chloride was introduced at mixing, for example, as calcium chloride, approximately 90% may form harmless complexes leaving only 10% as free chloride ions. If, on the other hand, sea water or de-icing salts penetrate the surface of the concrete, the ratio free to combined chloride may be 50:50.

The corrosive effect of chlorides is significantly affected therefore by the presence of free chlorides. The effects of chlorides are classified in terms of risk of corrosion because in certain conditions even low levels of chloride may pose some risk. The permissible level of chloride added at mixing specified in BS 8110 is 0.4% by weight of cement. For pre-stressed concrete the level is lower at 0.06%. The overall effect of reinforcement corrosion caused by chlorides must therefore be considered with the depth of reinforcement and the depth of carbonation.

There are two methods by which chlorides can be the cause of corrosion in reinforced concrete:-

- ❖ As cast-in chloride, usually calcium chloride added at the time of mixing or from sea-dredged aggregates.
- ❖ As ingressed chloride, ie by the penetration of the outer surface of the concrete from de-icing salts.

Chloride induced corrosion results in localised breakdown of the passive film rather than the widespread deterioration that occurs with carbonation. The result is rapid corrosion of the metal at the anode leading to the formation of a 'pit' in the bar surface and significant loss of

Materials and defects

cross sectional area. This is known as 'pitting corrosion'. Occasionally a bar may be completely eaten through.

Chloride induced reinforcement may occur even in apparently benign conditions where the concrete quality appears to be satisfactory. Even if there is poor oxygen supply reinforcement corrosion may still take place. Failure of reinforcement may therefore occur without any visual sign of cracking or spalling.

All aggregates used commonly for concrete mixing contain a background level of chlorides usually less than 0.06% by weight of cement. The use of calcium chloride as an accelerating additive at the time of mixing was popular during the 1950s and 1960s. It was used in precasting yards to speed up the re-use of expensive moulds and was used on site during cold weather to increase the rate of gain in strength. The use of calcium chloride was banned in 1977. The presence of calcium chloride cast-in within the mix usually attracts a chloride level significantly greater than 0.4% by weight of cement. Ingressed chlorides through the outer surface of the concrete are variable in nature. However the use of de-icing salts on, for example, external staircases and balconies is popular and may result in localised high concentrations of chlorides in excess of 1.0% by weight of cement. Concentrations of ingressed chlorides on the top surface of a car park deck may typically occur up to 3 or 4%.

The presence of calcium chloride is further exaggerated by the presence of deep carbonation. Carbonation releases combined chlorides into solution to form free chloride ions, thus increasing the likelihood of corrosion. For this reason many properties built approximately 20-30 years ago may only now start causing problems.

Risk assessment

Guidance on assessing the risk of reinforcement corrosion is provided by the BRE in Digest 444 Part 2. Here the risk of corrosion for structures of various ages is presented in the range negligible, to extremely high risk. Factors effecting the risk assessment are either a dry or damp environment, the depth of carbonation and of course the level of chlorides present.

Repair

The successful repair of chloride induced corrosion is notoriously difficult because of the tendency for new corrosion cells to form at the boundary of the repair. This mechanism is called 'incipient anode effect' and should be minimised by removing, wherever possible, all concrete with significant chloride contamination. In recent years the introduction of proprietary sacrificial zinc anodes embedded within the patch repair and attached to the reinforcement can help to reduce this effect. For high levels of chlorides and long-term protection this may not be sufficient.

For heavily chloride contaminated structures, particularly car parks, the only tried and tested long term solution is cathodic protection. The cost and complexity of installing cathodic protection is not usually warranted within building structures. A variation of cathodic protection is desalination – a short-term process using higher current densities than cathodic protection. Migrating corrosion inhibitors have found some success; these are penetrating surface coatings applied under strict conditions.

Materials and defects

Corrosion of metals

The electro-chemical series

Metal	Chemical symbol	Normal electrode potential (volts)
"Noble" or Cathodic, ie, protected end		
Gold	Au	+1.42
Platinum	Pt	+1.20
Silver	Ag	+0.80
Mercury	Hg	+0.80
Copper	Cu	+0.345
Lead	Pb	-0.125
Tin	Sn	-0.135
Nickel	Ni	-0.24
Cadmium	Cd	-0.40
Iron	Fe	-0.44
Chromium	Cr	-0.71
Zinc	Zn	-0.76
Aluminium	Al	-1.66
Magnesium	Mg	-2.38
Sodium	Na	-2.71
Potassium	K	-2.92
Lithium	Li	-3.02

"Base" or Anodic, ie corroded end

The further apart two metals appear in the table the more actively will the two metals react when placed in contact in a slightly acid aqueous solution.

Bi-metallic corrosion

Bi-metallic or galvanic corrosion is experienced when two dissimilar materials are in electrical contact and are bridged by an electrolyte. The electrolyte could be water containing salt, acid or a combustion product. The electrolytic cells comprise a series of positive anodes and negative cathodes between which current flows and at which electrochemical reactions take place. The degree to which a metal is subject to this form of attack is determined by the difference in voltage potential between two metals, the amount of moisture present, the relative areas of contact, the corrodent concerned and whether either or both of the metals have naturally occurring oxide films.

If it is necessary to use dissimilar materials they should be isolated with washers of a non-conductive material, for example, neoprene, PTFE, SRBF etc. Painting of the contact surfaces with bitumen is an alternative, but less reliable solution.

Effects of bi-metallic combinations

For cladding supports, aluminium alloys are often employed. When aluminium and stainless steel are in contact, there is the potential for corrosion to occur. The extent of corrosion will depend upon the respective sizes of the two metal components.

For example large aluminium fitting/small stainless steel bolt: little corrosion.

Materials and defects

Building defects

- Common defects in commercial properties
- Common defects in residential properties
- Problem areas with 1960s buildings
- Defects in concrete
- Fungi and timber infestation in the UK
- Rising damp
- Rising groundwater

Materials and defects

Common defects in commercial properties

In order to make a pre-acquisition building survey as useful as possible, surveyors should focus on how different types of buildings will be used and may be altered. Some of the issues to consider are set out below:

Shops/units

- Look for evidence of structural changes, such as removal of stairwells or load bearing partitions, to increase retail floor area.
- Check rear servicing for goods with clear access ways.
- Consider whether uniformly distributed floor loads are adequate. Historically, these may only be about 2.5 kN/m^2, whereas the current requirements may be for a minimum of 4.0 kN/m^2.
- There may be a basement, even though it is currently sealed.
- Check there is a proper arrangement for storage of refuse.

Offices

- Consider the likelihood that deleterious materials are present, as investors usually expect properties to be free of such materials.
- Investigate uniformly distributed floor and raised floor loadings.
- Consider how easily floors may be sub-divided into cellular offices or separate suites. The zoning of building services and structural/cladding/ceiling grids will be particularly important.
- Horizontal and vertical distribution routes for services need consideration.
- Consider the health and safety regulations for building occupiers.

Warehouses and industrial buildings

- Land with poor bearing capacity may result in differential settlement, affecting floor slabs, drainage runs and hard landscaping.
- Portal frame design may include an allowance for loadings imposed by tenant's services.
- Look for signs of damage and unevenness to floor slabs and enquire about uniformly distributed floor point loadings.
- Impact damage, particularly at low level to internal walls and to roller shutter opening reveals.
- Clear internal heights that are usually a minimum of 5.5 metres.

Materials and defects

Common defects in residential properties

Common defects in residential properties can be best categorised by the approximate age of the building, although this is only a rough rule of thumb.

Pre 1900

- Failed/lack of damp proof course
- Poor ventilation to suspended timber ground floors and roof spaces, leading to timber deterioration
- Poor fitting/ operating sash windows requiring overhaul/repair
- Damp penetration on solid brick walls
- Poor repairs to roofs, and valley/parapet gutters
- Roof spread due to recovering with concrete tiles
- Lack of restraint to flank walls giving rise to cracking of front and rear elevations and bulging or instability of flank
- Settlement of internal partitions
- Poor alterations to loadbearing (trussed) partitions
- Failure of brick arches and timber backing lintels above doors and windows - decay of timber lintols or bressumer beams
- Book end effect on terraced properties, where movement in a terrace can lead to the pushing out of flank walls
- Poor internal plaster on lath/plaster partitions
- Insect and/or fungal infestation especially in built-in timbers
- Defects in rainwater goods

1900 - 1939

- Wall tie failure in cavity brickwork
- Breakdown of render finishes to external walls
- Corrosion to roof tile nails
- Corrosion to metal windows and rot in timber windows
- Lead water mains
- Outdated electric services
- Corroded rainwater goods

Post-1939

- Failure of flat roof membranes
- Missing/inadequate wall ties to cavity walls
- Poor workmanship in structure ie the use of reinforced concrete in exposed positions such as lintels, with inadequate cover over reinforcement bars
- Presence of deleterious materials ie asbestos, used as insulant, in flue pipes, fire protection, or general building board

Materials and defects

- ❖ Outdated services
- ❖ Lack of bracing to trussed rafter roofs leading to lateral buckling of trusses and possible lack of restraint to gable walls

All residential properties, no matter what age they are, can suffer from defects relating to lack of maintenance, or poorly carried out maintenance and, increasingly, poorly executed DIY and 'improvements' such as:

- ❖ over notching of joists in running of services and central heating pipework;
- ❖ removal of spine/structural partitions without adequate support;
- ❖ removal of chimney breasts without adequate support of stacks;
- ❖ poor replacement windows;
- ❖ poor quality joinery; and
- ❖ poor sound insulation.

Problem areas with 1960s buildings

The 1960s saw a frantic period of innovation and experimentation. The need for replacement housing following the Second World War gave rise to the development of numerous system types using innovative materials. Construction had started to drift toward an assembly process and away from traditional craft-based skills.

Some 1960s construction was appallingly bad, for example,, some high rise social housing, but other schemes were of a very high standard using good quality and durable materials. Many buildings are now facing or have undergone major refurbishment or change of use (for example, office to housing or hostel accommodation), but many examples of good and bad construction remain.

Rather than look at typical problems associated with each style of building, this list is intended to give a very brief introduction to a number of common or typical materials or construction faults. The list is not exhaustive.

Materials and defects

	ITEM ELEMENT OR MATERIAL	EFFECTS
1.	**Aluminum sash windows (O)** A common type of window was the vertical sliding sash. Instead of the vision glazing being held in a frame, the glass ran in aluminium tracks, with horizontal top and bottom frame members clipped onto the glass. Spring balances were used to hold the windows open.	By now, these windows will be very worn. Defects in the springs or breakage of the glass can lead to the ejection of an entire sash window – clearly a health and safety issue. Treat these windows with caution.
2.	**Asbestos (A)** Very common in 1960s buildings. Chrysotile (white) for some insulation boards, roof sheets, water tanks, cill boards etc. Artex, floor tiles, partition wall linings, fire doors etc may also have a content. Amosite (brown) used as insulating boards, fire protection or fire breaks, behind perimeter heaters, partitions etc. Crocodilite (blue) often paste applied friable material in boiler rooms, pipework, calorifiers, etc.	Major health risks depending on type, location, risk of disturbance. Deleterious material, detection, management and control are highly regulated. Ask for a copy of the asbestos register.
3.	**Asphalt roofs (O)** These could be of quite good quality and may have performed well if laid on a concrete deck.	The lack of insulation would have helped to reduce temperature ranges and so restrict thermal movements.
4.	**Calcium chloride concrete additive (A)** Often used by manufacturers of pre-cast elements or for concreting in cold weather. Enables rapid set and removal of moulds. Can also be found in brickwork mortar. Use of un-washed sea dredged aggregates may have led to chloride contamination.	Considered a deleterious material. Creates conditions of high electrical conductivity in the concrete with consequent high risk of corrosion of steel reinforcement. Very difficult to repair effectively. Look out for spalling concrete and very black, stained steel. Can result in severe pitting corrosion without disruption of the surface of the concrete. Tests should always be recommended. Since about 1978, this material should not have been used. In brickwork, risk of corrosion of wall ties. Exposure to de-icing salts, salt spray etc can also be very damaging. Watch out for coastal locations, car parks, road bridges etc.

Key to common building types –
these are indicative only
W = warehouse or industrial
O = Office or commercial developments
H = Housing
A = All types

Materials and defects

	ITEM ELEMENT OR MATERIAL	EFFECTS
5.	**Calcium silicate bricks (A)** A smooth, often creamy coloured brick made from lime, sand and flint. Small particles of flint can sometimes be seen in cut bricks or weathered surfaces. Can be mistaken for concrete bricks (see next page). Widespread use in 1960s and 1970s and still manufactured and gaining popularity again.	Prone to shrinkage (unlike clay bricks which expand after laying). If movement control joints are missed or badly spaced (which they often were) diagonal cracking can occur. Thermal or moisture cracking often visible at changes in the size of panels for example, long runs below windows coinciding with short sections between windows. Look out for thin bed cracks and wider cracks to vertical joints. Do not confuse with subsidence cracking or corrosion of steel sub frame. Use as a backing to clay brickwork likely to cause problems as a result of expansion of clay brick and contraction of calcium silicate brick.
6.	**Cold bridging and condensation (A)** Poor insulation standards led to problems with severe cold bridging particularly in housing where humidity levels are higher. Polystyrene insulation was sometimes used but this was usually no more than 25mm thick.	Watch out for cold bridging around balcony structures and pre cast lintels
7.	**Cold flat roof construction (A)** Little thought was given to vapour control or for that matter roof insulation. It was common to provide sealed flat roof construction with minimal insulation and sometimes a foil backed plasterboard ceiling lining. Ventilation to the roof void was often ignored. Built up felt roofs were often asbestos based, but had a life of no more than 15 years. For this reason most felt roofs would have been replaced by now.	Risk of condensation occurring, with subsequent risk of decay to roof decking or to structure.
8.	**Concrete (A)** Can be of mixed quality, sometimes poorly compacted and with lack of cover to steel reinforcement. Under codes, depth of cover for external work should have been circa 40mm.	Sometimes, poor durability, corrosion due to the effects of carbonation or chloride content. Tests should be recommended. Poor curing methods could mean lack of durability.

Materials and defects

	ITEM ELEMENT OR MATERIAL	EFFECTS
9.	**Concrete boot lintels (A)** Concrete lintels designed to have a projecting nib to support the outer leaf, and built only into the inner leaf, to provide a neat appearance externally.	Rotation of the lintel under eccentric load, creating diagonal cracking to the brickwork above the window. Other signs are opening of the bed joint immediately above lintels and splitting of the reveal brickwork immediately beneath. Once rotation has taken place, brickwork will tend to arch over the opening, thus relieving some of the load on the lintel. Cracks can then be repointed.
10.	**Concrete bricks (A)** Similar in appearance to calcium silicate (cs) brick but often used in dark brown, dark red or dark grey variants. Harder and coarser texture than	cs bricks. Suffer from similar shrinkage related problems.
11.	**Corrugated 'big six' asbestos cement sheet (W)** Often found on industrial buildings and warehouses well into the 1970s. Name given as a result of the 6-inch profile, but in fact big six was one of several different profiles of sheet. Often based on an asbestos content of around 12 to 15% chrysotile, (white asbestos) with profiled eaves and ridge pieces and hook bolt fastenings. By the end of the 1970s insulation was being added to the roof construction and this brought about problems of condensation (See note on 1970s buildings).	Obvious health risks from fibre release. Friable surface, and very fragile – never walk on such a covering without crawling boards. Corrosion of hook bolts will cause sheeting to split. Often coated with bitumen or rubber solutions as a remedial treatment. Be very cautious of the effectiveness of these treatments.
12.	**External ceramic tiling (O)** See mosaic tesserae. Tiles were often prism shaped or ridged in some way.	Similar problems to mosaic tesserae in terms of delamination of background materials.
13.	**Flat concrete slabs (plate floors) (O,W)** Fairly thin slabs with mushroom head thickening around column heads.	Very high shear stress around column head has been found to be cause of structural failure. Beware of flat slab car park construction – recent major collapse of car park of lift slab construction.

Materials and defects

	ITEM ELEMENT OR MATERIAL	EFFECTS
14.	**High Alumina Cement concrete (O,W)** Often used in pre-cast and pre-stressed work rather than insitu work. Mainly (but not exclusively) for roof and floor beams. X and I profiles were common. Some variants for pre-cast factory units (portals and purlins). Sometimes used to form an insitu stitch between two pre-cast beam members or column-to-beam connections. Developed high early strength. Can have a brownish tinge.	Looses strength with age. Susceptible to chemical attack in damp conditions and contact with gypsum plaster. Strength and durability assessments to be recommended. Since 1974 the material should not have been used in buildings. Considered to be a deleterious material. Several roof failures in the 1970s, but no known cases of flooring collapse.
15.	**Hollow clay pot floors (O)** Quite common during the 1960s. Concrete poured between pots and in the form of a topping. Sometimes screed could be structural. In other cases, non-structural screeds may have been removed to gain additional load or headroom.	Watch out for clay spacer tiles between the pots. These can conceal honeycombing of the concrete rib, lack of fire protection, durability or strength. Removal of tiles and Gunite repairs may be necessary.
16.	**Lack of movement control joints (A)** Cement masonry walls are prone to thermal and moisture movements. Cement mortar is less flexible than older lime mortars, and the stresses induced by thermal movement are relieved by cracking. Centres of joints depend on nature of brick, size and shape of panel etc. Lack of joints was common in 1960s buildings.	Oversailing of brickwork on dpc, particularly in warehouses. Also watch out for joints that have been filled with Flexcell impregnated board, as this is not very compressible and can reduce the benefit of a joint. Early sealants were oleo resinous and could lead to staining of adjoining surfaces. Hardening and embrittlement of joint sealants to be expected now.
17.	**Large panel buildings (H)** A number of different systems were constructed. Large panels formed the external enclosure and also supported pre cast floor planks or slabs. Connection details were made on site with wire hoops and in-situ work.	Risk of disproportionate collapse – see Ronan Point disaster. Following this, high rise blocks were strengthened. Some low-rise blocks may not have been checked. Poor quality control of structural connections led to weakness, poor fire stopping or corrosion risk. Possible lack of tying-in of precast components. See 'Tying-in' later.

Materials and defects

	ITEM ELEMENT OR MATERIAL	EFFECTS
18.	**Mineralite render (O)** A thin (2-5mm) coating of fine grained minerals with a textured surface. Often applied to exposed concrete columns and beams or in larger areas such as spandrel panels. Variety of colours available.	Beware of adhesion failure – can be widespread. Difficult to match repairs.
19.	**Mosaic tesserae (O)** A common finish comprising small (25mm square) ceramic or glass tiles applied to a render background. Often supplied in paper backed sheets of around 300x300mm to facilitate laying.	Adhesion failure of tiles leads to individual tiles falling off and scattering over a wide area. Obvious health and safety risks. More often than not however, it is the render background that fails, loosing adhesion to the concrete or brick substrate. This is potentially more serious as larger and heavier sections could collapse. A hammer survey is to be recommended to check for soundness and to identify hollow areas. Repairs are possible using vacuum injection techniques, but hacking off and repair of spalled areas can lead to peel back and cracking of adjoining surfaces and deterioration due to water ingress and freeze/thaw cycles.
20.	**Mosaic tesserae in overhead situations (O)** Often used as a soffit finish to projecting balconies, shopping mall covered ways and the like. Tiles would be bedded on render, possible applied over expanded metal lathing.	Watch out for corrosion of the metal lathing or fixing screws as these may not have been protected against corrosion. Timber fixing battens behind the lathing can also be a problem. In severe cases, large sections of render and tile finish can collapse. If water penetration is suspected recommend further intrusive investigation.
21.	**No fines concrete (H)** Used in the manufacture of large panels for housing and similar structures, intended to create slightly better insulation properties.	Very low level of resistance to carbonation hence risks of carbonation and corrosion.
22.	**Panel joints (O,H)** Panel joints in large panel systems often comprised a plastic or metal baffle sprung into grooves in the edge of each panel. To prevent leakage it was common to provide a tape back seal to the rear face of the joint.	Baffles may be missing or dislodged. Back seals often missed with consequent risk of water penetration. Flat roof abutments often dressed under the bottom edge of panels, which makes them very difficult to repair. Usual need to modify the drained joint into a face sealed joint.

Materials and defects

	ITEM ELEMENT OR MATERIAL	EFFECTS
23.	**Reconstituted stone (A)** Often used as window sills or window surrounds, string courses or other projecting features in all types of buildings. Sometimes fixed with ferrous cramps rather than phosphor bronze. Contain light reinforcement.	Propensity to carbonate fairly rapidly with the result that reinforcement corrodes causing the features to spall. Corrosion of cramps can lead to displacement of features such as projecting window surrounds.
24.	**Render backgrounds (A)** Used in conjunction with tile finishes and mosaic tesserae. Often very strong Portland cement based mixes were used.	Possible adhesion failures on concrete due to presence of traces of mould oil on the surface. Sometimes used water based bonding agents (giving a white milky appearance when render is removed) when there was a poor mechanical key. Later bonding agents were of SBR which were more durable. Inflexible renders, high vapour resistance and risk of water entrapment.
25.	**Sand-faced Fletton bricks (A)** These were a popular and cheap brick type manufactured by the London Brick Company near Peterborough. Often found in 1960s housing or industrial applications. Often, but not always, a red/pink colour with a heavy textured wire cut type of surface. Rear face of brick is smooth with colour bands or 'kiss marks' arising from the burning process.	No problem in sheltered applications, but bricks in exposed situations such as parapets, chimney stacks and free-standing walls where saturation is common are at risk of sulphate attack. The bricks have a very high sulphate content. When wet, soluble sulphates react with Portland cement bedding mortars causing the mortar to expand and so disrupt the brickwork. Once this occurs, the damage is terminal. Often rendered in mistaken belief that this will cure the problem, but this is a very short-lived solution and will only make matters worse.
26.	**Softwood joinery (A)** External joinery was often of poorly seasoned sapwood with a low life expectancy.	Very poor durability, especially glazing beads, sills and horizontal rails. Further decay where timber has been pieced in during repair.
27.	**Steel windows and cladding (O,W)** Typical single glazed windows were manufactured using a section known as W20 by Crittal Windows. Either casements or tilt and turn varieties. Larger curtain walled sections were manufactured by coupling window units together with galvanised steel tee bars.	By now, early windows may be paint bound or distorted. Ironmongery may be defective. Timber sub frames were common and may be decayed.

Materials and defects

	ITEM ELEMENT OR MATERIAL	EFFECTS
28.	**Stramit roof decking (A)** An insulation board often used as a roof decking. It comprises a rigid board about 50mm thickness of compressed straw sandwiched between two layers of building paper. The boards were about 1200x450 width and had a brown paper finish. Check in plant rooms, lift motor rooms, roof access housings and the like. Used in some domestic applications. The material is still made, but is used for packing.	The material had a very low resistance to water and would decay easily. For that reason it is less usual to find it nowadays. The boards had a grain and needed to be laid correctly, with support perpendicular to the grain. Failure to do this could lead to distortion of the board and subsequent 'wave' effects in the roof line. This in turn could stress the covering and cause failure. Saturated Stramit board would turn into a brown silage like mess. Beware of safety issues (risk of collapse) when walking on Stramit roofs.
29.	**Tying-in of precast concrete floor and roof slabs (O,H)** Prior to 1972 (CP110) tying-in was left to engineering judgement. There is a need to form a connection between wall structures and internal precast floor planks. This can be achieved with continuous metal straps or structural toppings to prevent planks from gradually moving apart.	Failure to tie-in properly can lead to the elevation gradually parting company from the floors. Evidenced by a series of parallel cracks in the floors, gradually increasing in severity higher up the building. If neglected, collapse could occur under accidental loads.
30.	**Vitrilite panels (O)** Used in conjunction with steel windows as above, these single glazed spandrel panels were made from annealed glass with a powder coating fused into the rear surface during manufacture.	Risk of failure due to heat build up in spandrel panels, bird strikes or mechanical damage from cradles. Water penetration can cause staining and deterioration of rear surfaces. Replacement panels no longer available and may have been made from painted glass with less life expectancy.
31.	**Wall ties (A)** Often wire butterfly ties. Thin steel sections and poor galvanising standards. Cavity walls were rarely insulated and cavity tray detailing may be poor.	Acute risk of corrosion and poor wire spacing. Wire ties can corrode away without disruption of brick elevations leaving the outer leaf vulnerable to failure in extreme weather conditions. Wall tie population and sample survey should be recommended in situations where risk is possible. Thicker steel ties still suffered from poor galvanising standards. Corrosion of these is more likely to result in extensive cracking and/or 'pagoda effect' in flank walls.

Materials and defects

	ITEM ELEMENT OR MATERIAL	EFFECTS
32.	**Woodwool as permanent shuttering (W,O)** Often used in basement carparks where additional insulation was required, or in some office buildings.	Risk of poor compaction of concrete during placing, or grout loss leading to honeycombing around re-bars. This could prejudice fire protection, durability or in extreme cases strength. Intrusive investigation required to determine if steel is covered properly.
33.	**Woodwool slab roof decks** Often with galvanised steel tongue and grooved edge strips and with a pre-screeded finish, or an applied finish reinforced with chicken wire. Size about 1200 x 450 or 600mm. (See above for use in permanent shuttering). The material offered some thermal insulation qualities. Often used in plant room roofs, access housings and the like.	Reasonably durable and contrary to popular belief does not degrade rapidly when wet. However, failure of screed is probable during re-roofing operations, leading to need to renew the deck.
34	**Reinforced Autoclaved Aerated Planks (RAAC)** RAAC is characterised by low density aerated concrete with a cellular structure. The material is easily broken. Reinforcement is coated with cement latex or bitumen for protection. Often used for roof decks. Commonly known by trade names of Durox or Sipporex.	In pre-1980 buildings there is often excessive defection and associated mid-span transverse cracking. There is no evidence to suggest that RAAC planks pose a hazard to building users.

Materials and defects

Defects in concrete

The failure of concrete is often revealed by cracking, spalling and corroded reinforcement. While the outward symptoms of a number of faults may appear similar, repair methods must be based on a sound analysis of the cause.

Cracking can be caused before hardening due to workmanship problems or after hardening due to physical, chemical, thermal or structural effects.

The alkaline nature of concrete protects steel reinforcement against corrosion, a feature termed 'passivity'. A reduction in the level of alkalinity gives a risk of corrosion in steelwork provided water and oxygen are present.

Fault	Reasons	Results
Carbonation	Carbonation is generally of concern on exposed concrete surfaces and will take place very slowly with high quality dense concrete. Features such as poorly compacted concrete, cracks or other fissures in the surface and the type of aggregates used could all affect the rate of carbonation.	A reduction in passivity of concrete coupled with water and oxygen can lead to corrosion of steel reinforcement and subsequent spalling of concrete. Lack of sufficient cover to reinforcement may allow steel to fall within the carbonated layer and become at risk. It has been found that the depth of carbonation is roughly proportional to the square root of the time $d=k\sqrt{t}$ (d=depth, k=constant, t=time). The constant will vary with the properties of concrete. By examining the age of the concrete and the depth of carbonation it is possible to make a rough prediction of the depth of carbonation at a future date.
Rust stains	An indication of reinforcement corrosion but avoid confusion with staining ferrous sulphide inclusions in the aggregate or rusting of small diameter wires. Chlorine attack is difficult to deal with effectively and needs careful diagnosis (see later). Chloride induced corrosion will often result in dark stains around corroded reinforcement.	Although unsightly, staining such as this is unlikely to indicate reinforcement corrosion unless cracking is visible.

Materials and defects

Fault	Reasons	Results
Chloride attack		See separate section on chlorides on page 143
Sulphate attack	Sulphates are present in varying levels in many substances including gypsum plaster, certain aggregates and sub-soils.	Possible disturbance to foundations and walls due to expansion of floor slabs. Loss of strength, material becomes friable. Rate of carbonation increases, expansive reaction occurs in the concrete.
Alkali aggregate reactions principally Alkali Silica reaction	As a result of chemical interaction between alkaline fluids in concrete and reactive minerals in certain types of aggregates, a calcium alkali silicate gel is formed. This gel takes in water and expands.	Relatively rare fault in UK structures. In non reinforced concrete this cracking often illustrates a random pattern of fine, almost invisible cracks bounded by some larger cracks. This cracking is easily confused with shrinkage cracking or even frost attack. Gel can sometimes be seen on the surface of the concrete, possibly coupled with spalling lenses.
Aggregate reactions	Alkali carbonate reaction and alkali silicate reactions have similar problems, but are much less common in the UK.	In reinforced concrete, cracks tend to run parallel with reinforcing bars or prestressing tendons. In particularly severe cases, gel may be visible. Microscopic examination of concrete is the only sure way of identifying attack. Although rare, ASR is often only found in structures exposed to water.

Fungi and timber infestation in the UK

Fungi and moulds

Fungi live on dead organic material and have a natural role to play in the breakdown of dead organic material, which includes timber used in buildings. Most timber is too dry for fungal growth but timber decay can occur if the moisture content (MC) is increased.

Sporophores (fruiting bodies) of fungi produce millions of microscopic spores, which germinate to produce root-like hypha, growing and branching out producing a mass of mycelium. Suitable conditions are a food source, sufficient temperature, oxygen and moisture. The hyphae attack the wood cells and produce enzymes to break down the wood components. Sporophores, often the first indication of a problem, break out and begin the cycle again.

Wood rotting fungi, often termed brown or white rot, cause a change of colour, loss of strength and weight, splitting and cracking and surface

Materials and defects

mycelium to appear on affected timber. Brown rot attacks the cellulose component of the cell and wood develops cuboidal cracking. White rot attacks both cellulose and lignin in the cell wall but does not cause cross grain cracking. Wood rotting fungi familiarly is classified as dry and wet rot. Serpula lacrymans is the only true dry rot – there are many wet rots. Serpula lacrymans can decay timber at a much lower MC than any of the wet rot fungi and can penetrate masonry/brickwork and behind plaster.

The accompanying table should aid identification of various fungi. Dry rot can destroy timber with a 20% MC although 30-40% MC is the optimum. 20°C is the optimum temperature for dry rot to flourish. Wet rots require an MC of at least 30% with an optimum of 45-60%.

Various superficial coloured moulds, often seen affecting timber to buildings, are easily removed, but indicate a MC that might permit more serious fungal decay. Lowering the ambient humidity by increasing ventilation is generally the best course of action. The presence of non-wood rotting fungi also suggests conditions suitable for dry or wet rot.

Treatment

There are many specialist timber treatment companies who will carry out surveys, analyse samples and guarantee any eradication work they undertake.

Briefly, the traditional specification for treatment of dry rot will include the following:

- ❖ identify cause(s) of dampness and effect cure;
- ❖ cut out timber to 0.5m beyond decayed wood, remove and burn;
- ❖ hack off plaster, rendering and remove skirtings, architraves and other joinery from area to be treated to 1.0m beyond infection;
- ❖ remove surface mycelium from exposed masonry and wire brush;
- ❖ surface spray exposed masonry with fungicidal wall solution at manufacturer's recommended rate of application;
- ❖ consideration should be given to irrigating masonry, although it is unlikely that this treatment will be as effective as one might hope;
- ❖ replacement timber to be treated to BS 5268, 1977 and thereafter, treat with preservative to BS 5707; and
- ❖ existing timber to be cleaned and sprayed with organic solvent preservative and further treated by application of timber paste.

Treatment of wet rot is similar, although affected masonry need only be isolated from the source of the dampness. In some circumstances chemical treatments can be minimised and the outbreak controlled by environmental manipulation, but this may not always be practicable. Where chemicals are used, request specialists to provide COSHH assessments. For some time there has been considerable concern regarding chemical treatments. Approval for use of pentachlorophenol has been withdrawn and should no longer be used. Tributyltin oxide is also effectively banned. Permitted fungicides include zinc acypetacs and tri (hexylene glycol) bioborate. All specified treatments should display an HSE number.

Manufacturers' instructions should be closely followed and it is essential that:

- ❖ operatives wear protective clothing;
- ❖ washing facilities are available on site;
- ❖ there is no smoking;
- ❖ pilot lights to heating and cooking equipment are extinguished;

Materials and defects

- a fire extinguisher is provided;
- the treated area is adequately ventilated and nobody sleeps in the treated area for a minimum of 72 hours after treatment; and
- water tanks should be covered for 14 days after treatment.

In recent years there has been a trend towards less intrusive forms of repair. If the source of moisture can be removed, the fungus will die. Thus, by managing the building environment the problem of decay can be minimised, and the extent of disruption and intrusive surgery kept to a minimum.

The importance of safety measures cannot be over emphasised and it is vital that current safety legislation is understood and complied with, including the Control of Pesticides Regulations Act 1986 and Control of Substances Hazardous to Health Regulations Act 1988.

Under the Wildlife and Countryside Act 1981, it is an offence to spray roof spaces that may harbour bats without the approval of the Nature Conservancy Council.

Insect infestation

In this country wood boring beetles are the major group of timber attacking insects. The life cycle of the various beetle species is similar; adult female beetles lay eggs in cracks or the end grain of the timber, larvae emerge from the eggs and feed on the wood leaving frass in the tunnels they bore. The larvae enter the pupal stage and soon after emerge as adults leaving flight holes in infested timber. Some species cause only superficial damage.

Species that attack timber in use are more of a concern than those that only attack timber during seasoning or which attack standing trees or freshly felled logs. The MC of the timber is generally not important but insect infestation generally develops more rapidly on decayed or wet wood. Some insects only attack timber containing fungi. In most cases replacement of attacked timber is not necessary.

Examples of most common wood boring beetle species are as follows:-

- Common furniture beetle: very common, estimated that up to 80% of houses over 40 years old in rural areas are affected. Infestation often in damp areas of house, for example, beneath WC. Flight holes 1.5-2.0mm diameter. Adult beetles emerge May-September.
- Death-watch beetle: infestation uncommon, often found in ancient buildings and therefore more expensive to eradicate. Confined to south and central parts of England and Wales. Attacks elm, chestnut and oak. Adult beetles emerge in Spring through flight holes up to 3mm wide. Presence indicates fungal attack.
- Bark borers: found in timber where bark not completely removed. Larvae confined to bark areas and hence damage caused is superficial.
- House longhorn beetle: only found in Surrey, Berkshire and Hampshire. Regulations require new timber to be treated prior to use.
- Wood boring weevils: several species – only attack partially decayed timber, cause considerable damage.
- Powder post beetle: few flight holes, convert timber to powder leaving veneer of 'sound' timber.

There are many reference books available to aid identification. Briefly, treatment should include replacement of severely attacked timber with pre-treated timber, cleaning of all surfaces by vacuum cleaning and brushing, exposing area to be treated, and application under pressure of suitable insecticide such as permethrin or cypermethrin.

Materials and defects

Fungi identification guide

TYPE	USUALLY FOUND	EFFECT ON TIMBER
Wood rotting fungi		
Dry rot Serpula cuboidal	Inside buildings, mines, boats – never attacks timber outside	Large cuboidal cracking (brown rot)
Wet rots Conrophora Puteana (cellar fungus)	Most common of wet rots in buildings. Associated with serious leaks – fouled plumbing etc. Also decays exterior	Cuboidal cracking – small cubes (Brown rot). May leave thin veneer of sound timber. Affected wood becomes dark brown
Fibrioporia Vaillanti (Mine or Pore Fungus)	Associated with water leaks. Most common species of poria group	Cubodial cracking – large cubes (brown rot) Affected wood darkens
Phellinus Contiguous	Decay of external joinery (softwood)	Timber becomes soft. (A white rot) Wood becomes fibrous
Phellinus Megaloporous	Attacks oak heartwood. Presence often associated with death-watch beetle	
Corioius Versicolor (Polystictus)	Most common white rot decay of external hardwood	No splitting or decay but much weight loss
Lentinus Lepideus (Stag's Horn fungus)	Rare, but sometimes in flat roofs	Cuboidal cracking Darkens woods. Wood feels sticky

Non-wood rotting fungi

Peziza (Elf-Cup)	Occurs on saturated masonry or plaster, internally and externally. Associated with leaks	

Moulds

Aspergillus Penicillium Pullularia	Almost any damp surface in humid conditions	Superficial – easily removed

Materials and defects

MYCELIUM	FRUITING BODY	CONDITIONS FOR GROWTH
Cotton wool like if damp. Greyish white with purple/yellow and lilac patches if dry	Reddish brown centre, white margins. Flat plate or bracket shape Possibly red spore dust nearby	Timber MC 20-40% (slightly damp) Temperature 0-26°C
Brown branching strands on wood and masonry or brickwork. Usually not in daylight areas	Rarely found inside. Flat plate-like. Greenish brown centre, yellow margin. Knobbly surface	Timber MC 45-60% (very damp) Temperature -30°C-+40°C
Strands flexible when dry. White	Plate shaped – white pores. Rare	Timber MC 45-60% (very damp) Temperature up to 35°C
Light brown masses	Plate-like with pores. Dull brown	Timber MC 22%+ Temperature 0-31°C
Yellow	Large, plate-like, hard. Various browns in colour	Timber MC 20-35% Temperature 20-35°C
Rarely seen	Up to 25mm across. Hairy ringed zones to pores to underside	
Soft whitish needle shaped crystals on surface	Some resemble stags horns, others are inverted mushrooms on stalk – brown	Timber MC 26-44% Temperature 25-37°C
	Buff coloured and fleshy. Distinctive	
Like coconut matting	Toadstool – white head – spores released in black ink type liquid. Microscopic but spores show up as various colours: black, green, white, brown, yellow, pink	Very humid conditions

Materials and defects

Rising damp

Research by the BRE and others suggests that rising dampness is often misdiagnosed by surveyors and so called damp specialists, with the result that costly and unnecessary remedial treatments are specified.

In many cases, diagnosis is undertaken by means of electrical resistance or capacitance meters, but these can give very misleading and unreliable results in materials other than timber. Surveyors should not diagnose rising damp without first having undertaken a proper study. When rising dampness is suspected, do not automatically recommend a specialist inspection – more often than not the specialist will use exactly the same resistance equipment to make his diagnosis. A more reliable method has been prepared by the BRE and may be found in BRE Digest 245.

Symptoms

Typically these may include:

- ❖ damp patches;
- ❖ peeling and blistering of wall finishes;
- ❖ a tide mark 1m or so above floor level;
- ❖ sulphate action;
- ❖ corrosion of metals for example, edge beads;
- ❖ musty smells;
- ❖ condensation; and
- ❖ rotting of timber.

The above symptoms do not of themselves indicate the cause of dampness. Common causes could be lateral rain penetration, condensation or entrapped moisture. High external ground levels, bridging of damp proof courses, defective rainwater goods and the like should all be self-evident and could give rise to similar symptoms.

Soluble salts are present in many building materials. The salts can be dissolved and moved to the surface of the element as evaporation takes place. Hygroscopic salts (typically, nitrates and chlorides from groundwater) can be present in some materials. These salts absorb moisture from the atmosphere, and can in certain circumstances cause extensive staining and disruption of finishes.

Other possible sources of salt contamination include chemical spillage, splashing from road salt etc.

Equilibrium moisture content

Many building materials absorb moisture, and when exposed to damp air will attain an equilibrium moisture content. This hygroscopic moisture content (HMC) will vary according to relative humidity. Typical relationship curves can be plotted for different materials, and although these can only establish general indications it is possible to compare readings from different materials in the construction of a wall. For example at 75%RH the MC of yellow pine would be 13% while the MC of lime mortar would be 2% and 0.5% in brick.

Some materials possess an HMC of as much as 5% without the introduction of salts from external sources. This figure should be regarded as an appropriate threshold as to whether or not remedial action is likely to be required.

Materials and defects

Measurement of moisture content

Resistance or capacitance meters can give misleading results.

The presence of soluble salts on the surface of a wall will cause an electrical resistance meter to indicate a high reading, even if the wall were otherwise dry. Deep wall probes may give a more accurate picture, but will still be affected by soluble salts, as these are generally highly conductive.

A Speedy Moisture Meter will give a much more accurate reading of MC in all materials. (Resistance meters are usually calibrated for use in timber and can give an approximation of MC in that material).

The Speedy meter comprises an aluminium flask fitted with a pressure gauge and a removable lid. Using a 9mm drill on slow speed, a sample of dust is taken from the brick, mortar or plaster. The sample is weighed and placed into the flask. A small quantity of carbide is then added and the flask sealed. Moisture in the sample reacts with the carbide to form acetylene gas. The pressure of that gas is then read off the pressure gauge, which is calibrated to read %MC. With care, the meter can give a very accurate reading, comparable with laboratory kiln dried tests.

Rising damp

Rising dampness within a wall is in a sensitive equilibrium. There must be a supply of water at the base of a wall and the height to which that water will rise depends upon the pore structure, the brick, plaster or other finish. Water will also evaporate from the surface of the wall at a rate dependant upon temperature and humidity.

During wet weather, evaporation may decrease and ground water tables may rise, giving rise to an increase in the severity of the dampness. The reverse may happen during dry spells, and evaporation will be increased by central heating.

Soluble salts derived from groundwater or building materials will complicate the situation. Salts will increase the surface tension of the water and so draw it further up a wall. Furthermore, as evaporation occurs, stronger salt solutions are drawn towards the surface and may eventually crystallise out. This process reduces the amount of evaporation and so may raise the height of the dampness. The soluble salts are often hygroscopic and absorb moisture from the atmosphere. If this occurs, the situation will appear worse during wet weather and better during dry.

As noted above, the presence of hygroscopic salts does not necessarily indicate rising dampness.

Diagnosis of rising damp

BRE Digest 245 sets out a method of diagnosis. The method involves drilling samples from the wall to measure both their moisture content and hygroscopicity (HMC). Samples are taken from mortar joints from 10mm to a depth of 80mm every two or three courses from floor level up to a level beyond that which damp is suspected. While the carbide meter can be used to measure moisture content (MC) it will be necessary to send samples to a laboratory to measure HMC – to see if the samples could have absorbed the quantity of moisture found from the atmosphere.

By subtracting HMC from total MC, it is possible to determine the value of 'excess' moisture, which could result from capillary action or water from other sources. The comparison of HMC and MC gives an indication as to which is controlling the dampness at any position. If MC is greater than HMC, then moisture is coming from some other source. If the reverse applies, then moisture is coming from the air. Plotting the results graphically can then assist in gaining an accurate picture of what is happening.

Surface damage arising from hygroscopic salts can be significant. The HMC of contaminated wallpaper or plaster can be as much as 20%. It follows therefore that contaminated plaster will need to be removed. BS 6576 deals with this subject in more detail.

Materials and defects

Rising groundwater

During the latter part of this century, the level of ground water beneath major UK cities has risen rapidly, leading to increasing concern that huge costs could be incurred from damage to buildings and infrastructure if preventative measures are not taken. The problem stems from the city-centre industries that populated the areas during the industrial revolution and their demand for water. Beneath London this led to a reduction in the level of up to 90 metres. The usage has declined significantly since the late 1960s as the industries relocated and without this extraction levels have recovered, rising by 1.5 metres a year initially and by as much as 3 metres a year recently. By the late 1990s, water levels in central London had recovered by 35 metres, close to the 1900 level.

It was therefore realised that action to minimise the damage was needed urgently to allow time for planning and implementation. A group was formed from interested parties and in March 1999 it announced that a five-stage plan had been developed to safeguard London from the effects of rising groundwater. The plan involved controlled increased abstraction from 50 or more existing and new boreholes, with the amount that could be used for drinking water maximised and the remainder to be used for industrial or agricultural processes. The five phases of the strategy are:-

- ❖ Utilise four existing licensed water supply boreholes on the outskirts of London.
- ❖ Equip three proven borehole sites near central London with the latest filtration technology to provide further water for drinking purposes.
- ❖ Encourage the use of existing and new private boreholes, in and around central London where the water quality is lower, for non-potable uses.
- ❖ Identify locations for new control boreholes in central London, initially to be pumped to waste until end uses can be determined.
- ❖ New control boreholes in outer London to intercept the flow of water towards the central basin.

The plan has the backing of government and is being managed by Thames Water, working in conjunction with the Environment Agency. Although the threat to London is the most immediate, other cities are at risk and it is likely that similar strategies will be implemented as required.

Testing

- Chemical and physical testing requirements
- Non-destructive testing

Materials and defects

Chemical and physical testing requirements

The following are some of the more common requirements for specific tests, although it is better to discuss the quantity or type of sample required with the testing laboratory before sampling to ensure that a suitable and representative sample is provided.

Testing laboratories should be provided with an outline of the location of the sample and the nature of the element inspected. This is often critical in evaluating results and the possible level of risk involved.

Primary Rules

- ❖ avoid contamination of sample
- ❖ label clearly
- ❖ inform laboratory of purpose of test
- ❖ consider health and safety aspects when dealing with hazardous substances; seek expert advice first
- ❖ for comparison, 10 grammes = a sugar cube.

Chloride ion content

About 30-50 grams (enough to fill a 35mm film canister) of sample material is required. Drilled dust is preferred obtained using a 10mm percussion drill bit. Discard first 5mm depth of material to avoid contamination from paint, plaster or other surface effects. Take samples from two adjacent holes drilled to the depth of reinforcement. As chloride levels will vary, many samples should be taken for analysis. For chloride profiling to identify ingressed chlorides then take incremental samples at depths typically 5-25, 25-45 and 45-65mm.

Carbonation

Best carried out in-situ. Drill two 10mm diameter holes and break out concrete in between. Treat freshly exposed concrete with phenolphthalein. Concrete will turn pink if un-carbonated. This is not a precise test and carbonation should be recorded to the nearest 5mm.

If you find the depth of carbonation by testing and you know the age of the building, then the rate of carbonation increase in the future can be predicted according to the following formula: $d = k \sqrt{t}$ where d = depth of carbonation, k is a constant and t = age of concrete (years).

Refer to BRE Digest 405.

High Alumina Cement (HAC)

Dust samples taken, as for calcium chloride ion, to determine aluminium content and assess 'proof negative' the presence of HAC. Differential Thermal Analysis (DTA) required on lump samples for 'proof positive' confirmation. Phenolphthalein test for carbonation is not applicable to HAC. It is necessary therefore to remove a lump sample of sufficient size to enable a thin slide to be made for petrographic laboratory analysis. This test is very expensive.

Refer to BRE Digest 392.

Sulphates in concrete

As for calcium chloride ion (30g sample should be sufficient for chloride sulphate and HAC).

Materials and defects

Plaster/mortar

To determine mix proportions including, for mortar, cement content. Take dust or preferably solid sample as for chloride ion content.

Asbestos

Air monitoring and bulk sampling by specialists only. Samples must be taken with due regard to health and safety. Seal loose material after sampling. Sample of loose insulation may be taken with corer. About a thumbnail size sample is sufficient for analysis. Laboratory to be UKAS accredited. Seal remaining material after sampling. Laboratory to report on asbestos type, % content and density.

Ground water

Approximately 1,000ml of water should be sufficient to determine chemical signature of ingressed water. A similar quantity of mains tap water should also be taken simultaneously so that the results may be compared. Results of laboratory analysis sometimes are not conclusive.

Leaks may also be identified by 'sounding' by placing listening rods on pipe valves, covers etc. This work is done by specialists often when quiet at night.

Decayed timber

Provide sample large enough to indicate pattern of cracking and/or mycelium growth. Powder material is not sufficient except in cases of insect attack, where frass may give indication of type of beetle.

Non-destructive tests

See BS 1881: Part 201, 'Guide to the use of non-destructive methods of test for hardened concrete' and information supplied below.

Non-destructive testing

There are a range of non-destructive testing (NDT) techniques which may avoid the requirement for large scale costly and disruptive opening-up. Often the findings of non-destructive testing survey may allow opening-up to be targeted in specific locations. Non-destructive testing techniques are particularly useful to establish construction details and the condition of the structure and fabric of a building, in particular relating to building defect analysis.

A summary of the more common NDT techniques are as follows:-

Impulse radar

Radar is a widely accepted non-destructive technique for establishing details of the construction and condition. Radar is effective through most building materials including concrete, asphalt, brick, stone and also through soil. Defects may be mapped without damage or disruption to surface finishes.

Radar is an echo sounding technique where pulses of radio energy are transmitted into the structure by an antennae moved over the surface. Where material boundaries are encountered the different electrical properties reflect part of the energy back to the surface where it is detected by a receiver. Sampling is rapid and collected data effectively forms a continuous cross section enabling rapid assessment of thickness, arrangement and condition over large areas.

Thermography

Infrared thermography is a non-destructive testing method involving precise measurements of surface radiation to reveal changes in thermal performance caused by hidden changes in the construction or in a physical condition. Recent improvements in imaging and processing technology has enabled improved and compact handheld thermographic cameras to undertake surveys of large areas relatively quickly.

Modern thermo-imaging cameras are able to detect variations in surface temperature as small as 0.10°C. A major limitation of this method is that sufficient heat differential is required between the inside and outside of the building fabric. Environmental conditions favourable for thermographic surveys are therefore restricted to, for example, cold still winter nights when heat flows from the inside to the outside of walls. A major benefit of this technique however is that a great deal of work can be conducted some distance from the subject.

Ultrasonic pulse velocity

Ultrasonic pulse velocity is a non-destructive technique used in testing a wide range of building materials to determine properties including compressive strength and to investigate defects, such as the presence of delamination and the depth of cracking. The method can be applied to materials such as concrete, ceramics, stone and timber. The main advantage of this method is identifying general changes in condition such as areas of weak concrete in a generally sound structure. The technique involves the measurement of the compression wave velocity from a transmitter to a receiver.

NDT investigations

Often one or more of the above techniques may be required. Information may be provided relating to:

- ❖ debonding and delamination;
- ❖ compaction and voidage;
- ❖ spalling or micro-cracking;
- ❖ general details of construction, material types and layers;
- ❖ moisture content;
- ❖ reinforcement or other embedded ferrous metal; and
- ❖ brickwork details and conditions.

It is rare that NDT techniques will themselves tell the whole story. It is essential that NDT is considered as an investigation tool as part of a wider engineering assessment.

Cladding

Materials and defects

- Curtain walling systems
- Mechanisms of water entry
- Glazing – windows and doors satisfying the Building Regulations
- Spontaneous glass fracturing

Materials and defects

Curtain walling systems

> Curtain walling is a weatherproof and self-supporting enclosure of windows and spandrel panels in a light framework which is not built up within the main structure of a building but is suspended right across its face, being held back to the structure at widely spaced points.

Types of curtain walling based on the erection process

Stick system

Stick construction comprises a grid of mullions and transoms into which the various types of infill can be fitted. Most of the assembly work is done on site. The advantages include relatively low cost and the ability to provide some dimensional adjustment. The disadvantage of the system is that it depends on site labour for some of its performance. It is not unusual to find up to 90% of systems failing initial waterproofing tests during erection.

Unitised or panellised

Unitised construction comprises a factory-assembled unit complete in all respects including glazing. The advantages are that pre-assembly under controlled conditions and rapid enclosure of the building removes the likelihood of failure. The disadvantages include the bulk of the modules and the difficulty of leaving small openings for site access, scaffolding, etc.

Combination systems

In the unit and mullion system, pre-assembled units are offered up to erected mullions. This system is used where the mullions are of a large cross section or area. The sizes of the units tend to be smaller than in the fully unitised wall. In the column span system there are three basic components: covers that clad the vertical structural elements; long spandrel panels that span between them; and pre-assembled glazing unit infills.

Types of curtain walling based on the mechanics of resisting rain penetration

- ❖ front sealed
- ❖ drained
- ❖ drained and ventilated
- ❖ pressure equalised, drained and ventilated
- ❖ rainscreen cladding

To accommodate or to resist rainwater entry, the above forms of curtain walling have been designed. There are limitations as to use and performance of each.

Early curtain walling systems tended to be face sealed; in other words they attempted to deal with the weather head-on. Water was not permitted to reach the inner part of the glazing but to run off the surface like a raincoat. Unfortunately it is extremely difficult to guarantee a totally watertight façade and if some water penetrates the structure there is unlikely to be a path for it to drain out again.

There are still several variations of face sealed systems available. Some of the new silicone glazed systems are fully face sealed. These systems comprise glass panels bonded in place using structural silicone glazing compounds. Structural rigidity is also provided by mechanical fixings on

Materials and defects

the surface of the glass. Front zipper cladding systems are also manufactured. These systems usually contain a large rubber gasket with a central, press-in segment or zip which, when pressed into place, forces the gasket out on to the surface of the glass. Again, this system does not normally have provision for water drainage.

Most cladding designers now accept that it is difficult to exclude water and therefore provision for a small amount of leakage can be made. Many systems comprise an assembly as follows:-

- ❖ A decorative cover plate (usually colour coated).
- ❖ Pressure plates incorporating either two narrow gaskets or a single 'eagle' gasket. May be termed a 'clamping plate'.
- ❖ A thermal break.
- ❖ The mullion or transom member.
- ❖ A further gasket (or back seal) behind the glass.

Some systems (for example, Schuco) also incorporate a foil-faced adhesive tape – a butyl tape – applied over the transom and mullion directly beneath the pressure plate to serve as a secondary line of defence.

Having accepted that some of the water will pass through the outer line of defence, namely the pressure plate, it is important to provide a drainage mechanism. Systems will either be mullion drained or transom drained. Small holes are formed in the pressure plate to enable water to drain off from the glazing rebate. For a drained system to work properly, it follows that there must be a clearly defined drainage path. Water draining from the transom and into the mullion will pass vertically down through the system to a drainage point, usually a small plastic baffle set into the mullion above a joint.

The problem with a simple drained system is that small amounts of moisture may be retained within the cladding members. This water could cause deterioration, particularly of the edge seals to the double glazed units.

The next variation of the basic system is therefore a drained and ventilated system. In this case additional, or enlarged holes are let into the pressure plates. Water is encouraged to drain out of the system but in addition small amounts of air are allowed to ventilate around the edge of the glass. So any small amounts of water that have entered the glazing system will evaporate.

While the designer of the drained and ventilated system accepts that some water will bypass the outer seals, it is nonetheless important that the pressure plate is fixed to the correct torque so that the outer gasket seal forms a good seal against the glass.

The back seal has a critical role in that it prevents any movements of air from the glazing rebate into the building. Any gaps or leaks in the back seal can lead to water being drawn into the building as a result of air pressure differentials.

Some of the cheaper cladding systems utilise single piece gaskets that are cut and mitred at the corners. In many cases workmanship during installation is lacking which means that the gaps at corners make the system vulnerable to water ingress. Because it is normally only possible to see the joints in the back gasket when the glass is removed, accurate diagnosis of problems with the backs can be difficult. A better system will utilise a single piece gasket with vulcanised corners to reduce the risk of a poorly cut mitre being formed during installation.

As noted above, air pressure differentials are often a source of moisture ingress into a cladding system. To combat this, designers have come up with a system of pressure equalisation.

Materials and defects

Consider the following diagram:

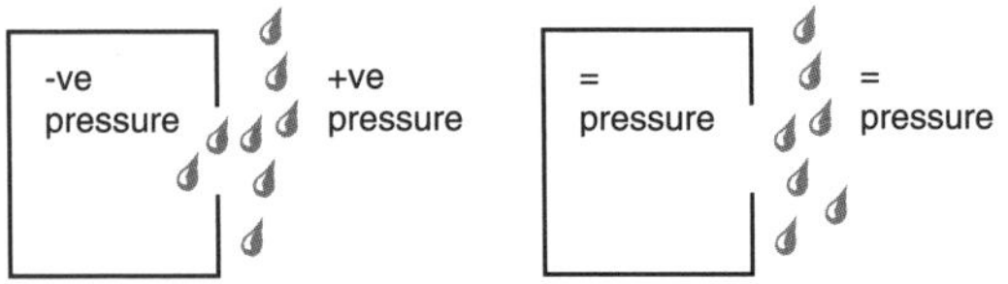

The rectangular box may be considered as the area around the glazing in a cladding system. If the pressure inside the box is less than the pressure outside, the water drop will be drawn in. If we can equalise the pressure in the box then the probability is that water will stay on the outside. In curtain walling terms we must provide a perfect seal around the rear of the window – the back seal – and must also provide a number of slots around the perimeter of the glass to enable pressure within the glazing rebate to equal that of the external air pressures very quickly.

It is very difficult to identify the differences between a drained and ventilated system and a pressure-equalised system. However, for pressure equalisation to work properly, the various zones of pressure must not be too large. Thus it is common to consider the area around one glass pane as one zone and therefore drainage must be made from the transom members and not from the mullions. In practice it is very difficult to provide a fully effective pressure equalisation system, and there is some doubt in the industry as to whether they are fully effective.

Testing

Many standard systems have been erected and tested in special test rigs. Taywood Engineering operates one of the few major testing rigs in the United Kingdom. Samples are tested against a number of standard tests covering wind resistance, air permeability, and water resistance, etc. Specifications for cladding will often contain certain design criteria, but for these to be meaningful the criteria must be both achievable and capable of being tested.

Laboratory testing often involves the use of an aeroplane engine to simulate fluctuations in pressure and extreme weather conditions.

Site testing is however much more difficult. It is possible to erect a pressure box on one side of a glass panel. By reducing the air pressure within the pressure box it is possible to simulate and to monitor performance of components in situ. However, even this system is inflexible and can only realistically be carried out in isolated areas. For this reason field tests for curtain walling have been designed. The American Aluminium Manufacturers Association originally designed the common test for in situ conditions but it has been adopted by CWCT.

Standard hose test to check water tightness

- ❖ The purpose of this test is to check the workmanship of the curtain wall as constructed.
- ❖ The check is to determine the resistance to water penetration of only those joints in the wall that are designed to remain permanently closed and watertight.
- ❖ The specifier shall designate areas of the wall to be tested.
- ❖ The first areas to be tested should be among the first areas of each type of curtain wall constructed on the project.
- ❖ These areas should be at least one structural bay wide and one storey in height, provided that all horizontal and vertical joints or other conditions where leakage could occur are included.

- ❖ The water shall be applied using a brass nozzle that will produce a solid cone of water droplets with a spread of 88 degrees. The nozzle shall be provided with a control valve and a pressure gauge between the valve and the nozzle. The water flow to the nozzle shall be adjusted to produce 22 ±2 litres per minute, when the water pressure at the nozzle inlet is 220 ± 20 kPa.

- ❖ With the water directed at the joint and perpendicular to the face of the wall, the nozzle shall be moved slowly back and forth above the joint, at a distance of 0.3 metres from it, for a period of five minutes for each 1.5 metres of joint while an observer on the indoor side of the wall, using a torch if necessary, checks for any leakage and notes where it occurs.

Cladding design: active and inactive systems

Early cladding systems were based on window technology. The initial thrust of design development was to improve their wind and rain-tightness through better sealing systems between panels, then drainage, ventilation and ultimately 'pressure equalisation' of joints to stop water penetrating and being pumped into the interior by wind pressure. Most recent curtain walling systems that the surveyor will encounter will either be drained and ventilated, or more likely, drained, ventilated and pressure-equalised. It can be difficult to distinguish which features apply solely from a visual inspection. In this case, reference to drawings and design details is required to avoid error.

Subsequent developments in the technology led to the separation of the wall into two constituents: the 'rainscreen' and the 'technical wall'. The rainscreen is the external decorative face of the building and carries most, if not all, of the responsibility for excluding rain. The technical wall supports the rainscreen from behind and provides vapour control and insulation. In some cases, the envelope is further elaborated with the introduction of external louvres or brises soleil to reduce solar gain, and access ways to enable maintenance and cleaning. These designs are increasingly complex and bring with them a whole range of performance and maintenance issues, but they are still 'passive'.

The active wall approach is becoming an increasingly common feature of advanced commercial buildings. The cladding system is in two transparent skins – an outer single or double glazed layer and an inner glazed layer. Solar gain though the composite system is controlled by way of blinds within the cavity. During the heating season, cool air from the building is drawn into the cavity and is preheated by the sun (or an additional heater) before being recirculated back into the space. During the cooling season, warm air from the cavity is drawn off and either discharged or used to power a heat pump to produce cooling.

These are exciting times in building construction. However, as with all developments, performance-in-use data lags behind. We do not yet know what problems will be experienced as the systems age, nor what will be the implications of rectifying them. This does give cause for some concern, as remedial works to the external envelope can be notoriously disruptive and costly. Clients commissioning or acquiring buildings with complex cladding systems need to satisfy themselves that all reasonable risk issues are fully considered and designed-out as far as it is practical to do so.

Surveyors need to gather and appraise as much information as is available and above all to understand how the cladding has been designed to work. In the process the following questions need to be answered:-

- ❖ How does the design life of the building compare with the life of the cladding?
- ❖ Is the system passive or active?
- ❖ What building physics are behind the design – what are the mechanisms of air and water entry?

Materials and defects

- ❖ What are the properties of materials used in fabrication and how compatible are they?
- ❖ How does the design of the cladding system relate to the design of the structure?
- ❖ What are the allowances for tolerance and movement?
- ❖ What were the test results when the building was commissioned (90% of systems fail on first attempt)?
- ❖ What is the maintenance and cleaning strategy?

Mechanisms of water entry

Ways rainwater can penetrate

- ❖ kinetic energy
- ❖ surface tension
- ❖ gravity
- ❖ capillarity
- ❖ pressure differentials
- ❖ any combination of these

Kinetic energy

This is the direct action of the wind carrying a droplet of rainwater with sufficient momentum to force it through a sealed joint. Prevention or drainage overcomes this. Prevention can be by baffles, by a durable seal or by a labyrinthine shape within the joint. Drainage collects the penetrating water and diverts it back to the outside.

Surface tension

This can cause water to adhere to and move across surfaces. It is guarded against by drip edges or throatings along leading edges, and horizontal surfaces should slope down and out. Connecting components can also have appropriate grooves or ridges.

Gravity

This can take water through open joints that lead inwards and downwards. Reversing the slope overcomes this.

Capillarity

This occurs in fine joints between wettable surfaces. It is only severe when other mechanisms persist – for example, wind-assisted capillarity. In metal components it is resolved by capillary breaks within the joint surfaces.

Pressure differential

This frequently is the main mechanism. It is overcome by maximising the outer deterrent and minimising the pressure differentials. This is achieved by self-contained (compartmentalised) air spaces behind the outer skin, which are well ventilated to the outside.

Materials and defects

Glazing - windows and doors satisfying the Building Regulations

The 2002 amendments to part L of the Building Regulations have brought about important changes in the specification of windows and doors. Essentially, compliance with the regulations will involve improving the insulation value of the frame (perhaps by reducing its size and increasing the extent of the better insulated glass panel) or by using more advanced glazing specifications such as low e coatings.

Specifically, measures could include the following:-

- ❖ Increase the width of the thermal break.
- ❖ Utilise an alternative type of thermal break.
- ❖ Provide alternative gasket design.
- ❖ Increase the air gap in the double glazed unit.
- ❖ Provide a gas filling to the glass.
- ❖ Provide an alternative spacer (warm edge technology).
- ❖ Utilise low e glass (either hard or soft coatings).
- ❖ Triple glazing.
- ❖ Dual double glazing.

Increase the width of the thermal break

The width of the thermal break will have a significant impact on the thermal performance of the frame. Current (pre April 2002) designs are often in the order of 12mm or less, but widths of 36mm are also realistic. Even with low e glass and an argon gas filling, a standard 4mm thermal break in a metal window will not meet the requirements.

Utilise an alternative type of thermal break

There are broadly two types of thermal break: those comprising a strip or 'web' of nylon 66 (polyamide) or glass fibre reinforced polypropelyne inserted into locating channels in the aluminium sections.

There is also those of resin which is poured into a single extrusion. When the resin has set the aluminium is then cut out.

By replacing a polyamide break with one made of polyurethane, one can reduce the u-value from say 3.95 W/m≤K down to 3.72 W/m≤K. However, polyurethane is not capable of accommodating the wider break designs.

Provide alternative gasket design

The thickness and positioning of gaskets, particularly those types, which also compartmentalise the glazing rebate, can influence thermal performance to a comparable degree.

Increase the air gap in the double glazed unit.

Many domestic DGU's are of 12mm width, with some timber window frames being only able to accommodate a 6mm air gap with two panes of 4mm glass. By increasing the width of the gap to 16mm, thermal performance is improved. Beyond 16mm there is little thermal benefit to be gained.

Provide a gas filling to the glass

Conventional DGU's are air filled, although the substitution of air with argon or krypton gas will lead to a marked improvement:-

- ❖ Air filled low e double glazed unit U=1.9 W/m≤K.

Materials and defects

- ❖ Argon filled low e double glazed unit U=1.6 W/m²K.
- ❖ Krypton filled low e double glazed unit U=1.35 W/m²K.

If the gas filled unit is well made it is possible that the gas will remain contained for many years. Estimates suggest that after manufacture, about 90% of the volume will be the specified gas with the remaining being air. A good unit should not loose more than a further 5% of gas over the next 25 years. However, there are no established test methods for establishing how much gas has been lost in service. The quality and effective life of gas filled units is therefore unknown at the present time.

Provide an alternative spacer (warm edge technology)

A high performance DGU such as those described above would not function as effectively as it could because standard aluminium spacer bars have a high thermal conductivity. By replacing the spacer bars with systems of low conductivity, it is thought that improvements of as much as 10% can be achieved. However, there are some concerns over the long term durability of these units. Additional types of warm edge spacers such as silicone foam cannot be tested under the methods described in prEN 1279-2 and this could restrict their use in the future.

Utilise low e glass (either hard or soft coatings)

Low e or low emissivity glass is made by coating one surface of a pane with a special metal coating. The treated pane is then encapsulated into a DGU, with the treated pane on the inner or room side. The coating reduces the amount of light that can pass through in the infra red spectrum, thus more heat is trapped on the inside of a room.

There are two types of low e glass:
- ❖ Pyrolitic hard coat; and
- ❖ Post applied soft coat.

Pyrolitic hard coat (for example, Pilkington k and SGG EKO Plus) is applied while the glass is being made. Thus the coating is very resilient, meaning that the glass can be toughened, cut and assembled without the need to remove the coating at the spacer bar contact point.

Soft coat (for example, SGG Planitherm) is applied post manufacture. Very high thermal properties will be dependent upon the type of coating used, but most soft coats cannot be toughened and are easily damaged. DGU's must be assembled within a short time of being cut. For these reasons, soft coatings tend to be more expensive, but some manufacturers will claim that they can provide improved thermal performance over hard coatings.

Triple glazing

A triple glazed window using float glass is not as effective as a DGU with Low e glass, and is much thicker resulting in heavier frame sections. However, a triple glazed unit with two low e panes can achieve 1.0 W/m≤K, while if the voids are filled with krypton, 0.7 W/m²K can be achieved.

Dual double glazing

As its name implies, this involves setting two DGU's within a frame. However, the system would be very heavy with associated handling and cost implications. Thermal bridging at openings might be reduced because of the increased width of these systems.

Materials and defects

Spontaneous glass fracturing

Nickel Sulphide is one of several chemical contaminants that can occur during the manufacture of glass. There is some debate as to its origin, but it is thought that it is due to the mix of nickel and sulphate impurities within the glass batch materials, the fuels or even the furnace equipment, and this creates polycrystalline spheres which vary from microscopic to 2mm in diameter.

All glass has some of these inclusions present; they are impossible to eliminate entirely and therefore they are not considered a product defect.

In untreated (annealed) glass they are not a problem. But when glass is heat treated (toughened or tempered), the inclusions are modified into a state which transforms with temperature and time and which is accompanied by an increase in volume.

In a majority of cases this has little effect but dependant on size and proximity to the centre of the pane where the forces are greatest, this can eventually cause the glass to break.

There is a theory that for an initial period of approximately one year after manufacture there are relatively few breakages. After this, the number increases for up to several years, thereafter decreasing in frequency. There have been reported incidences where fractures have occurred more than 20 years after the installation of glass.

Panes in external situations are at greater risk. However, there have been a small number of cases where spontaneous breakage has occurred in internal glazing, remote from external influences, for example, panels to a staircase balustrade or an internal partition.

Although the safety risks are very small, because of the risk of falling debris, some companies will not recommend or supply and glaze toughened glass for sloped overhead applications where it will be used either as a single pane or as the inner leaf of a sealed unit.

In 'heat strengthened' glass, Nickel Sulphide inclusion is not generally regarded as a source of fracture. The difference between this and toughened glass is the rate of cooling. In the former this is less rapid, reducing surface compressive strength and making it much less susceptible to the transformation of Nickel Sulphide inclusions. Offering 5.5 times the strength of annealed glass, in many circumstances it is a useful replacement for tempered glass. However, it is not a suitable substitute where safety glass is required.

'Heat soaking' is a quality controlled process which gives increased reassurance against the presence of critical Nickel Sulphide inclusions by subjecting the glass panels to accelerated elevated temperatures to stimulate the transformation of the crystals and thus initiate immediate failure. It is thought that this process identifies 90% or more glass, which might have subsequently failed after installation. The heat soaking process could be used either as a sampling method or as an additional treatment, which in the case of clear toughened glass could add up to 20% to the cost.

Heat soaking does not change any of the physical properties of toughened glass and therefore there is no means of distinguishing whether or not this process has been carried out.

Identification

When toughened glass is broken, the tensile stress is spread out from the source causing the pane to crack into small fragments (dicing). These fragments tend to be slightly wedge-shaped, emanating from the source of the fracture and are often held into position wedged against the frame due to their increased volume.

Materials and defects

If the fracture is as a result of expansion of the Nickel Sulphide inclusion, those fragments immediately adjacent are more hexagonal and at the epicentre of the breakage the two larger particles form a distinctive butterfly shape linked by a central straight line crack. If large enough, the inclusion may be seen in the form of a black spec, or its presence may be confirmed by optical microscopy.

When carrying out an investigation, all possible causes of failure should be considered, including poor glazing tolerances and insufficient allowance for subsequent movement of the frame and any supporting structures. Possible causes may include deflection or rusting of steel frame, shrinkage of concrete frame, thermal movement, normal air pressures and even sonic booms. If the fracture is a result of impact or of local point loading, there should be evidence of local crushing.

The chances of installing a toughened glass pane, which may later fail due to the expansion of Nickel Sulphide inclusions, are very small.

Where it is essential, for reasons of accessibility or safety, that the pane should not fail, alternative forms of glass should be considered.

Testing

British Glass, (tel: 01142 686201) will test a failed sample to confirm the existence or otherwise of Nickel Sulphide inclusions. Samples should be sent intact with the epicentre protected with clear film.

Conservation and the environment

- Contaminated land
- Energy conservation
- Environment and specification
- Radon
- BREEAM

Conservation and the environment

Contaminated land

Environmental due diligence is now common place in property transactions. The presence of contaminated land can adversely affect site value and rental income and can hinder transactions if not managed properly.

With an increasing move away from greenfield development, it is no longer possible for the majority of investors or tenants to avoid owning or occupying some contaminated land, whether as a city centre property that has had a variety of past uses, or a new out of town development constructed on an old industrial site.

Documentation

Documentation is the key to maximising value and rental income and ensuring a smooth transaction. Documentation should be in place to demonstrate the following:-

- The site condition has been adequately assessed by phase I desk based research and phase II intrusive surveys and there are no significant information gaps (see below for an explanation of phase I and phase II).
- Where contamination has been identified, it does not represent a risk to the existing or proposed use of the site.
- Contamination is not migrating off-site within groundwater.
- Contamination does not represent a significant risk to groundwater and surface water resources.
- Contamination does not represent a risk of Regulatory Authority action.
- Contamination does not represent a risk of third party action (for example, from adjoining land owners).

Phase I

'Phase I' means an assessment that is based on background research and sometimes a site inspection/walkover and will normally include a review of:

- historical site uses;
- current site uses;
- environmental sensitivity; and
- Regulatory Authority Records.

Phase II

'Phase II' means an assessment that is based on an intrusive physical investigation of the site, and usually includes chemical analysis or monitoring of soil, water and land gas.

If a Phase I report identifies a significant issue, then it may be necessary to conduct a Phase II intrusive investigation, to gain an understanding of whether contamination is actually present and whether it is likely to represent a significant risk.

The Phase II investigation should be designed to address the specific issues raised by the Phase I report. The Phase II investigation should look for the range of contaminants highlighted by the Phase I report as possibly being present (for example, petrol on a petrol station, landfill gas on a landfill site etc).

The majority of sites undergoing development will require Phase II

Conservation and the environment

investigations, particularly where the previous use was industrial. Phase II contamination investigations can often be combined with geotechnical investigations for foundation design. Sites that are not being developed may require Phase II investigations (for example, during transactions), depending on the specific circumstances.

Assessment

In order to allow the client to make an informed decision, contaminated land surveys should include an assessment of:

- ❖ whether contamination is present (or is likely to be present), and the types of contaminants;
- ❖ the likelihood of contamination representing a risk to the current or proposed use and to other receptors such as rivers and groundwater; and
- ❖ the likelihood of regulatory or third party action against the site.

Environmental insurance

Environmental insurance is increasingly being used in property transactions to cover risks associated with contaminated land. Environmental insurance can be obtained directly from specialist underwriters, or through an insurance broker. A broker will normally obtain quotations from a number of different underwriters, in order to negotiate the best insurance cover and premium for a client.

The most common type of environmental insurance covers regulatory and third party claims due to land contamination.

When taking out environmental insurance, one of the most important aspects is to understand what circumstances the policy will not cover. Furthermore, as with many insurance policies, environmental insurance policies are often written in a language that can only be easily understood by an insurance expert.

Common exclusions from policies are:

- ❖ 'known' contamination;
- ❖ business disruption costs;
- ❖ remediation costs on change in use; and
- ❖ loss in value.

Key contaminated land legislation

The government believes that the planning process provides the best means of remediating sites. Nevertheless, new contaminated land legislation was introduced in 2001 in the UK (as Part IIA of the Environmental Protection Act 1990) intended to deal with problem contaminated sites that are not being developed and which would therefore not be dealt with under planning.

Definition of contaminated land under Part IIA

For the purposes of the Part IIA legislation 'Contaminated Land' means land where substances are present in, on or under, the land that are causing, or are likely to cause, significant harm or pollution of controlled waters.

Many sites that are contaminated will not fall within the definition and will not be classified as 'Contaminated Land' under Part IIA. However, such contamination could still have implications to owners and occupiers, for

Conservation and the environment

example, in terms of affecting the saleability and marketability of a site and may still require a contaminated land assessment.

Who is liable under Part IIA?

- **Remediation notices** are served in the first instance on 'Class A' persons ie polluters or knowing permitters of contamination. If these parties cannot be found, the current owner or occupier may be responsible – 'Class B' persons. Several parties may be implicated.
- **Sellers can pass on liability** where there are **payments for remediation**, with an explicit statement in the sale contract that a purchaser is being paid to clean up land or that the purchase price is being reduced to reflect the contaminated state of the land.
- **Sellers/landlords can also pass on liability** by **selling with information** – giving the purchaser (or tenant under a long lease) the necessary information to identify contamination before buying. Where transactions have occurred since 1990 between large commercial organisations, the granting of permission by the seller for the buyer to carry out its own investigations as to the condition of the land, should normally indicate that the buyer had the necessary information.

How the Part IIA regime works

The Part IIA regime is enforced by local authorities, or by the Environment Agency/Scottish Environment Protection Agency in the case of Special Sites (the most seriously contaminated sites which include military land, land used for oil refining and nuclear sites).

Once land is identified as being contaminated, local authorities have to achieve clean-up by serving remediation notices. (For Special Sites, responsibility for ensuring clean-up rests with the Environment Agency or Scottish Environment Protection Agency).

There is a three-month consultation period before a remediation notice is served, to encourage voluntary remediation (except in emergencies).

As of October 2001, 19 sites in the UK have been determined as being contaminated within the meaning of the legislation.

Progress with Part IIA legislation

As of the end of March 2002, local authorities (in England) have designated 33 sites as contaminated within the Part IIA definition. Voluntary remediation has commenced at seven sites (undertaken by either the original polluter, the site owner or occupier, or the authority).

- Most of the designated sites are relatively small (<5 hectares).
- Fuel/oil storage and waste industries are common causes of contamination at the designated sites.

Energy conservation

Some of the ways in which energy efficiency can be enhanced are outlined below:-

- **Greater thermal insulation.** This is one of the most cost-effective ways of increasing a building's energy efficiency. With new buildings, obviously the starting point is to meet the standards of thermal insulation set out in the current

Conservation and the environment

building regulations. However, there is a strong argument for further increasing thermal insulation and the level chosen will depend upon a variety of factors including the pay-back period required, the user's pattern of occupation and whether environmental concern outweighs strictly financial considerations. Our existing buildings offer great scope to increase insulation and this can be carried out during refurbishment or maintenance periods when better use can be made of access equipment and other site overheads.

- ❖ **Efficient lighting.** In commercial buildings, lighting costs usually run between 40% and 50% of total energy costs. Highly efficient lighting that can significantly reduce running costs is now available for both new and existing installations.

- ❖ **Efficient services.** These require preventative maintenance to ensure that all plant is operating at maximum efficiency. When plant requires renewal, consideration should be given to alternatives, such as condensing boilers or combined heat and power plant, or even whether the plant is required at all; many are re-evaluating the need for air conditioning.

- ❖ **Building management systems.** These monitor and control all service installations. Due to recent technological advances, these systems are becoming less expensive and can therefore be installed on smaller properties. Even where a full BMS is inappropriate, simple systems are available which control single services such as lighting management systems.

- ❖ **Using locally sourced materials.** This minimises energy consumption when transporting materials to site.

- ❖ **Building materials.** Those with a long life expectancy imply energy efficiency because they make good use of resources. The manufacture of building materials entails energy consumption and this varies widely depending on the product. Unfortunately, there is insufficient research into the energy consumed during the manufacture of building materials to allow choices to be made with a great deal of confidence. In the meantime, a useful rule-of-thumb is that the greater the degree of processing or manufacture, the greater the energy consumed.

- ❖ **Making use of solar gain.** Even with their humblest dwellings, our ancestors frequently designed their buildings to make use of solar gain. Generally this entails the use of large areas of glazing on southern elevations and minimal openings to the north. Increasing numbers of building designers are re-interpreting these techniques.

- ❖ **The use of soft landscaping.** Trees and other planting can conserve energy in buildings by minimising heat gains and losses. A screen of deciduous planting at the south of a building will filter strong summer sun but will allow for natural heat gain from weaker winter sun. Soft landscaping also has an important part to play in influencing the micro-climate around buildings by reducing wind speeds.

- ❖ **Out-of-town schemes.** Many of these schemes are built with high levels of thermal insulation and efficient building services. Nevertheless, because of the fuel used in transporting building users from their homes, the whole scheme may be very inefficient in energy terms.

Increased energy efficiency is available at little extra cost – it is just a matter of adopting an environmental train of thought. Increased awareness of these issues by players in the property market and pressure of legislation should be seen as an opportunity to create more energy efficient buildings.

Conservation and the environment

Environment and specification

The environment does not lend itself to simple right-or-wrong selection criteria of materials. However, careful specification does play a part in a complete design and operating philosophy for environmentally responsible building. There are, of course, environmental aspects to the use of all materials but for reasons of space only two model specification clauses are included here; timber and chlorofluorocarbons.

Timber

All references to timber contained within the specification are to be obtained exclusively from sustainable sources. The contractor is to provide evidence, by a supplier's certificate and labelling for each consignment delivered to site, that the timber is from such a source.

The certificate and label should include the following information:

- ❖ the species and country of origin;
- ❖ the name of the concession or plantation;
- ❖ a copy of the forestry policy; and
- ❖ shipping documents confirming the source.

The Good Wood Seal of Approval by Friends of the Earth would suffice. Information on appropriate suppliers can be obtained from the Timber Trades Federation, Friends of the Earth and the International Timber Trades Organisation.

CFCs, HCFCs and halons

In line with current good practice, the following clauses relating to chlorofluorocarbons (CFCs) hydrochlorofluorocarbons (HCFCs) and halons, their removal and restrictions on use, shall be deemed to have been allowed for in any tender or estimate offered.

Existing plant

Where CFCs or HCFCs are identified as being contained within existing air conditioning or refrigeration plant, the contract administrator is to be informed immediately and instructions obtained.

Where the plant is scheduled, or instructed subsequently, for removal then under no circumstances is the gas to be dumped by venting into the atmosphere. The gas is to be collected for recovery/destruction by a specialist firm.

The contract administrator is to be informed in writing of the specialist undertaking the works, the date for removal and be provided with a copy of the recovery/destruction certificate.

New plant

The contractor shall receive and transmit to the contract administrator documentary evidence from suppliers, sub-contractors and designers of all new installations that no new or re-used plant contains refrigerants with an ozone depletion potential of more than 0.06 (or with any ozone depleting potential). Furthermore, compounds with the lowest possible ozone depleting potential are to be selected where there is a choice.

As an alternative, consideration can be given to the use of absorption chillers or ammonia chillers.

Conservation and the environment

Halon fire fighting systems

The general requirements for de-commissioning the systems shall be the same as for CFCs and HCFCs in plant.

If the system is to be tested, then compressed air or some other non-ozone depleting gas shall be used.

If the system is to be recharged, a leak detection system should be installed. However, the preference is for some other form of fire extinguishing system wherever possible. Depending on the circumstances options include inert gases (Inergen, Argonite, etc), carbon dioxide and water fog/mist systems.

Hand-held fire extinguishers

All fire extinguishers on site supplied by the contractor, or specified to be supplied in the Schedule of Works section shall be either powder, foam, carbon dioxide, water spray or some other type without the use of Halon.

Additional measures

CFCs and HCFCs are used in the manufacture of a variety of other products including insulating materials, carpets, furnishings and aerosol sprays. They must not be used unless specifically instructed by the contract administrator.

Radon

Radon is a radioactive gas that occurs naturally in the earth. It has no taste, smell or colour and detectors have to be used to test its presence. When concentration levels rise within buildings, it can pose a serious risk to health.

Radon occurs when uranium decays and becomes radium. When radium decays, it becomes radon. Uranium is found in small quantities in all soil and rocks. It can also be found in building materials derived from rocks.

Radon rises from the soil into the air. Outdoors, radon is diluted and the risk it poses is negligible. Problems occur when it enters enclosed spaces, such as buildings, where concentration levels can build up.

Radon is everywhere but usually in insignificant, variable quantities. There are some areas in the UK where geographical effects result in higher levels.

The National Radiological Protection Board (NRPB) has produced maps of radon affected areas, which include Derbyshire, Devon, Cornwall, Northamptonshire, Somerset, Yorkshire and the Lake District in England, Grampian and Highland regions in Scotland and County Down and Armagh in Northern Ireland, and Bristol, Newport and West Cardiff in Wales.

The NRPB has set threshold levels for both commercial and residential properties. Detectors should be installed to ascertain the level of radon present.

If the readings exceed the threshold level, remedial works will be required in order to reduce the effects of radon. They offer a search service and report, which specifies whether a property is in a radon affected area and the probability that radon will exceed the action level. However just because a property lies within one of these areas it does not necessarily mean that it will have a problem with radon. The British Geological survey also states whether radon is likely within a particular area as part of its Address Linked Geological Inventory.

Conservation and the environment

Identification of radon

Radon levels vary appreciably with time, so prolonged measurements are required for reliable results. Short measurements can be misleading, low or alarmingly high. The government recommends that people in affected areas test their property for a period of three months using passive monitors in order to provide a reliable estimate of the average radon level. Passive monitors are easy to use, inexpensive and available from DEFRA. A good rule of thumb is one detector per 100m^2 of floor area. In larger open planned work areas such as a production area in a factory or an open plan office, the number of detectors may be reduced to one per 500m^2 of floor area.

Remedial action for high radon levels can be quite straightforward. The best approach is to prevent radon entering the building from the ground by altering the balance of pressure between the inside and outside.

This can be achieved by carrying out the following:-

- Install a small sump pump below the floor and connect to a low power fan in order to extract the air and reduce the pressure under the floor. This is known as an active sump (a passive sump relies on natural forces to drive air through the system). To minimise inconvenience, the sump may be outside the building with a pipe through the wall. A sump is a reliable and effective remedy that can reduce radon levels by at least a factor of ten. Multiple sumps can be used in large buildings.

- Improved ventilation under suspended timber and concrete floors. New airbricks are installed, sometimes together with a fan. This system again limits the amount of radon entering the property.

- Increase the pressure in the building by blowing air (called positive pressurisation) from the roof space with a small fan. Best results are in buildings with low natural ventilation. This is a reliable remedy that can at least halve radon levels. Secondary benefits may include a reduction of other indoor pollutants such as carbon dioxide, reduced condensation and a 'fresher' indoor environment.

- Alternatively, one may seal ducts, joints and cracks in the floors although this is rarely effective by itself and always laborious. It is helpful to close large openings when a sump is used.

- Ventilation (other than positive ventilation).

- Install a membrane barrier. This is very difficult to successfully achieve in an existing building.

For domestic properties the DETR recommend six main ways to reduce indoor radon levels to significantly below the action level of 200 Bq/m^3.

Remember that it is the average exposure to radon that matters. Short exposure at high levels is not important if over the long term your average exposure is low. This means that you should have time to plan for the solution that is best for the client, property and the radon level. But having found the best solution, it should be implemented as soon as practical.

It is best to stop radon entering a house or, if that is not possible, to try and remove it if it gets in. Solutions are similar to those recommended for commercial properties, but with subtle differences:

- Install a radon sump pump. The system limits the amount of radon that enters the house and for a typical house it is by far the most effective method. Modern sumps are often constructed from the side of the house so there is no disruption inside.

- Improve ventilation under suspended timber floors.

Conservation and the environment

- Use positive ventilation. This system is designed to change the air pressure in the house by blowing air in from the loft level. The system both dilutes the radon to acceptable levels and stops some of it getting in.
- Seal cracks and gaps in the floors.
- Change the way the house is ventilated. This solution is only suitable in quite special cases and has drawbacks.
- Install a membrane barrier.

When action has been taken and radon levels reduced, it is recommended to arrange routine checks on fans and other equipment. Periodic measurements of radon should also be made for overall assurance that levels remain low. These guidelines apply to all types of buildings.

Radon and the Building Regulations

With the new understanding of radon risk, the government legislated that houses built since 1988 in parts of Devon and Cornwall and 1992 in parts of Somerset, Derbyshire and Northamptonshire had to have radon protection measures built in.

Two zones of risk were allowed for. First the primary zone (the area with the highest risk). Requirement C2 of Schedule 1 of the Building Regulations requires that each house has a radon proof area, together with other precautionary measures that can be upgraded if a risk shows high radon levels. In the secondary zone (where the risk is lower) only precautionary measures must be built in. If a house has precautionary measures, upgrading them (for example, adding a fan to a sump and pipe system) could solve the radon problem quickly and simply.

The BRE has published guidance on protective measures for new dwellings in support of the Building Regulations entitled BRE Report 'Radon: Guidance on Protective Measures for New Dwellings'. No equivalent advice has been prepared for commercial buildings, but it is recommended that similar measures will be applicable for non-domestic buildings.

BREEAM

The Building Research Environmental Assessment Method (BREEAM) is the world's most widely used system for assessing, reviewing and improving a range of environmental impacts associated with buildings.

Since its launch in 1990 BREEAM has been increasingly accepted in the UK construction and property sectors as offering best practice in environmental design and management. Buildings are assessed against performance criteria set by the BRE and awarded 'credits' based on their level of performance.

The building's performance is then rated as pass, good, very good or excellent.

BREEAM covers a range of building types: offices, homes (known as EcoHomes) and industrial units. Other building categories can be assessed using a bespoke version of BREEAM.

BREEAM 2002 was launched on 31 August 2002 by the Building Research Establishment to supersede BREEAM 1998 and all new assessments are carried out under this scheme.

BREEAM for Offices is updated every year to ensure best practice and relevance to changing standards and regulations. The challenge to achieve the highest rating has increased as targets are being continually raised.

Developers and designers can utilise BREEAM for a range of reasons including creation of better environments for people to work in, increased

Conservation and the environment

building efficiency, improved marketability, increased value and rentals and also as a checklist for comparing buildings.

Clients and developers can use BREEAM as a tool to define and specify the environmental and sustainability performance requirements of their buildings at the briefing stage. Agents can use the rating to promote the environmental credentials and benefits of a building to potential owners and tenants. Designers can use BREEAM as a method to improve the performance, environmental and sustainability aspects of buildings.

A BREEAM Office assessment of the building fabric and services is undertaken plus, as appropriate, the quality of the design and procurement and also management and operating procedures.

The scheme requires a commitment to a number of areas, listed below, that are reviewed by independent assessors who are trained and licensed by the BRE.

- Management: overall policy, site management via the Construction Confederation Considerate Constructors Scheme and procedural issues.
- Health and well being: both internal and external issues affecting occupants health.
- Energy efficiency including operational energy and carbon dioxide issues.
- Transport: carbon dioxide and location related factors.
- Water consumption and efficiency.
- Materials: environmental implications and life cycle impact.
- Land use regarding greenfield and brownfield sites.
- Ecology including enhancement of the site as well as ecological value conservation.
- Pollution of air and water

To achieve an excellent rating the most cost effective way is to address the main issues at the earliest point of the design process with input from the full project team.

A BREEAM assessor can be used to coordinate and collate input from the team and to track the development of ideas. The assessor can also give advice about BREEAM to the entire project team at the start of the project.

As the scheme progresses, the assessor can provide specialist advice on the specification of products to achieve particular BREEAM credits, undertake preliminary BREEAM assessments to assess the predicted rating and provide a sustainability report inclusion with submission for planning approval.

At completion a certificate is awarded and this can be used for promotional purposes.

Maintenance management

- Primary objectives
- A systematic approach
- Condition surveys
- 'Best value' in local authorities
- Sources of information in maintenance management

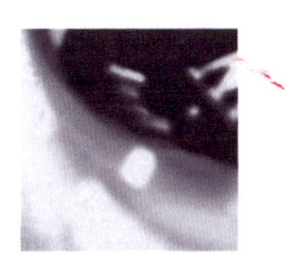

Maintenance management

Primary objectives

It is essential that property owners and managers allocate sufficient time and resources to maintenance. Managers need to be aware of the full extent of maintenance liabilities and how much money should be spent and when. Maintenance management requires a systematic approach to ensure high standards, value for money and management control.

Buildings comprise a number of elements. Their constituent materials and components have a range of life expectancies that in most cases will be shorter than the life of the building as a whole. Maintenance is therefore inevitable and arises from failure at this component level.

The primary objectives of maintenance are:-

Protection of the investment to ensure high utilisation of the building and its long life

Maintenance affects the profitability of a commercial organisation in a number of ways. First there are the direct costs of labour, plant, materials, and management. The second category is indirect where inadequate maintenance of buildings prevents the organisation from functioning properly.

Safeguarding the return on the investment

Poorly maintained commercial buildings fare unfavourably in the market when compared with their well maintained equivalents. The run down of an investment from lack of maintenance will discourage tenants from wishing to remain in occupation. This will have an effect on rent levels and encourage assignments and vacations on termination of leases.

The control of costs

Timely, planned maintenance enables efficient use of resources.

Establishing a safe working environment

The safety of the building and establishing a safe working environment must be the first priority of the maintenance manager.

Maintenance closes the gap between the actual state of the building and the acceptable standard. The acceptable standard will be dependent upon a number of factors, which may include:

- statutory requirements (health and safety);
- tenant or occupant satisfaction;
- minimising loss of production;
- morale of users, employers and customers; and
- public image.

A systematic approach

A systematic approach to maintenance management has a number of elements.

Policy

A building may be an asset to an investor or a resource to the user. The maintenance policy defines the objectives that maintenance of the building sets out to achieve. It is a dynamic concept, subject to change just as the plans and objectives of the user or organisation will change.

Maintenance management

Standards have to be defined. The use of 'normal standard' is inadequate. The acceptable range or performance of any element of a building will depend upon the relationship between each of its functional requirements and the use of the building as a whole.

The policy must therefore contain objective criteria to define what constitutes failure or non conformance in each category. The policy must identify those activities that are sensitive to the physical condition of the building and those building elements that play a significant role in providing the necessary conditions.

Once these components are identified, the acceptable delay time in correcting any failure can be assessed.

The statement of policy constitutes the brief for the maintenance manager. It should cover future requirements of the buildings, changes of use, statutory and legal conditions, maintenance cycles, required standards and acceptable response times for breakdown or failure.

The policy must be regularly reviewed and amended as necessary. Consideration of a maintenance policy often shows that there may be some conflict between parties with different interests, for example, between landlord and tenant.

Survey and data gathering

In order to measure how the buildings compare against the policy, and how to close the gap, it is essential to review all existing maintenance information and to undertake regular condition surveys. These identify the condition of the asset and record the status of the building at any one time. A condition survey has the specific purpose of:

- ❖ identifying maintenance needs;
- ❖ recording the priority of the need;
- ❖ recording proposed remedies and quantities of items requiring attention before the next survey; and
- ❖ predicting the scale of items requiring attention after the next survey.

Planning of work

Having agreed a policy for the organisation and identified work to be done, the first function of the maintenance manager is to formulate a maintenance plan. The objective of planning is to ensure that work is carried out with maximum economy. A key function of maintenance management is to achieve positive control over the work and to avoid overloading or inefficient utilisation of resources. Planned maintenance consists of preventive and corrective work.

Preventive maintenance

That which is carried out at predetermined intervals and is intended to reduce the probability of failure.

Corrective maintenance

That which is carried out after failure has occurred and is intended to restore an item to a state in which it can perform its required function.

Generally speaking, the most economic plan in direct cost terms would be the one that maximises preventive maintenance. However, by definition, this form of maintenance takes place before failure has occurred and some degree of useable life has been wasted. This wasted life has a value that can be costed and must be added to the direct cost of the maintenance resource used. The split of work between preventive and corrective maintenance is a management decision. The actual proportion of each will be determined by reference to the maintenance policy.

Maintenance management

Plans must take into account practical considerations and the availability of resources. They are prepared with different time horizons for different purposes:-

- ❖ Long term – for strategy, to establish general expenditure levels and to profile the maintenance demands of a property or portfolio over an extended time horizon.
- ❖ Medium term – addressing demands that are predicted within the next five years and to refine budgeting and the assessment of workload.
- ❖ Annual – for work and resource allocations.

Organising

The construction of an organisation and a control system capable of ensuring the implementation of the plan.

The main constituents of a maintenance organisation:-

- ❖ Resources – ie labour, materials, plant and management. These may be either directly employed or obtained through contracts with outside contractors.
- ❖ Administration – a staff structure for coordinating and directing resources.
- ❖ Work planning and control – the system for work, budget, cost and condition control. At its heart lies a documentation procedure, which ensures that an information base exists for effective decision making and that data essential for effective maintenance is transmitted around the system.

The documentation system will include a number of elements such as an asset register, a property information base, and preventive maintenance documentation together with a system for initiating and controlling works: works orders, work request forms, etc. These systems are generally computerised. There are a number of proprietary computer programs available. Such programs revolve around relational databases that can be tailored for strategic work, planning and budgeting right through to processing of day-to-day orders for reactive maintenance.

Procuring

A contract and procurement strategy must be established. The strategy will be driven by the nature of anticipated and known workloads and the need to achieve maximum efficiency of operation.

The workload may include the need to provide reactive maintenance cover for unforeseen or accidental repairs, 24 hour emergency call out cover, planned maintenance contracts and small items of maintenance work packaged into cost-effective contracts.

The decision about which contract and procurement strategy to adopt will depend on a number of factors including the geographic spread of the portfolio in question; the preferred type of model arrangements to be implemented; and the number and type of resources available to the maintenance manager.

Monitoring

Monitoring and auditing is essential to ensure that the maintenance management system is functioning properly and achieving the requirements of the policy statement and if not, to record deficiencies and initiate corrective action.

The maintenance manager needs to monitor the system and organisation to ensure that quality and value for money is being achieved. The audit is a post examination of maintenance work and procedures, not dissimilar to a financial audit. The audit can be condition, technical, systems or design based.

Maintenance management

The condition audit involves the checking of planned maintenance schedules against the actual condition of the building.

The aim of the technical audit is to examine a sample of maintenance activities to investigate how each task has been approached and dealt with.

The systems audit will analyse maintenance management practices. It will establish the true nature and extent of the database being used and how this is stored and accessed. It will look at overall policy and whether or not there are realistic plans and programmes in existence, at the method used for budget preparation, budget control, and for feedback of information to assist managers in future decision-making.

The design audit concentrates on the interaction between design and building performance. In most cases it is a question of lessons to be learned for the future.

Condition surveys

Property assets are extremely important to owners, investors and occupiers and therefore need to be properly maintained and managed (see also 'maintenance management'). Current research indicates that UK business fails to understand the importance of it's property assets, an important aspect of which is building condition which deteriorates with time, reducing value and utility.

Condition surveys represent an essential means of collecting data and ultimately reporting on the physical condition of a property or portfolio of properties, as well as addressing performance and statutory compliance issues if required.

The RICS Guidance Note defines a condition survey as follows:

> The collection of data about the condition of the building, part of a building, estate or portfolio assessing how that condition compares to a pre-determined standard, to identify any actions necessary to achieve that standard now; and maintain it there over a specified time horizon; the purpose being to support management decision making.

Condition surveys are of benefit to everyone involved in the ownership, occupation and investment in property. For owners, occupiers and managers, these surveys provide a database of information identifying maintenance and repair work as well as the time scale for undertaking that work allowing budgeting, assessment of compliance and strategic planning. For investors, it represents an audit to measure the condition of their asset, while for purchasers it assists in asset valuation by identifying repair and dilapidation liabilities. At strategic/policy making level, it will enable both a view of policy effectiveness, and allow implications of regulatory change to be established.

Format and content

A brief must be clearly defined and agreed. The brief should set out the number of properties to be inspected; for example, for portfolios of large numbers of similar types of property, a sample is often taken rather than surveying the full portfolio. It should be confirmed whether the surveys will comprise the internal and external building fabric, whether or not building services are to be included as well as infrastructure items such as roads and main services; hard and soft landscaping and boundaries; and extent and detail of statutory compliance items. Agreement should be reached on the depth of the survey and the approach to complex defects which will often need to be noted as requiring further investigation, as

Maintenance management

well as whether any specialist tests and surveys, such as drain surveys, form part of the requirement. The method of costing items should also be established.

The format and content of the reports will inevitably vary across clients and will relate to how they will aim to use and present the data, how they wish to prioritise the works identified, and what type of timescale the condition surveys are considering. Typically, condition surveys tend to consider the condition and works required over five to ten years but may address much longer periods.

The format and content of the surveys will also depend on how the data is to be collected and stored. With the availability of powerful computerised databases, and hand-held computer technology for gathering data, clients may not in fact require much in the way of traditionally written reports, while others would prefer to have a textual report with a basic chart or spreadsheet indicating building assets, building description, condition, priority of remedial works, along with a time frame and budget cost.

It is recommended that there should always be a written report summarising the findings of the survey in addition to confirming the limitations and the brief. It should include a summary of the results, possibly with easily understood graphics, as well as suitable cross reference to electronic files that may have been provided in the process.

Priorities

The classification of repair priorities will also vary but, typically, items affecting health and safety will have the highest priority, while works necessary to maintain an element in repair, or those maintaining civic/corporate standards may be less important. For example:-

- **Priority 1/serious/unavoidable** – Typically relating to health and safety compliance; items requiring immediate attention to avoid a major breakdown, serious hazard or critical deterioration leading to possible closure of the building.

- **Priority 2/poor/essential work** – Typically works that if neglected might lead to damage to the value of the property, and premature replacement.

- **Priority 3/fair/necessary work** – Where neglect might affect rental income, the condition of the element is less than adequate and defects need to be remedied within two/three years.

- **Priority 4/adequate/desirable** – Works necessary to maintain an element in repair but probably not in the immediate short term. Might include items that would improve working conditions or generate financial benefits such as energy conservation.

In order to avoid ambiguous interpretation of repair items, it is wise to keep the number of categories to a minimum. Whatever categories are adopted should be clearly defined with the survey team appropriately briefed and trained.

The subsequent use of data collected depends very much on how it is stored. It is essential to use spreadsheets or preferably databases that allow data to be filtered and sorted with enquiries run as appropriate to the beneficiaries of the surveys. Prioritisation of condition should, wherever possible, be undertaken using computers as this enables items to be evaluated with more than one factor, typically not only physical condition but also consequential effect on the asset and the building occupant, as well as management factors and policies.

Maintenance management

Frequency of surveys

It is unfortunate that often expensive condition survey exercises are undertaken providing what effectively results in a 'snapshot' of the condition at the time of the survey but thereafter the data is not used or updated. Clearly, in the case of a property transaction, this 'snapshot' assessment of repair liability is appropriate but to ensure cost effective long term maintenance of building assets, the condition data should be periodically updated. British Standard 8210: 1986 suggests in-depth surveys on a five-year programme, supplemented by two-yearly inspections on a more superficial basis. Clients may choose a different profile but generally the view is that, for maintenance purposes, intervals of a maximum of five years are acceptable. Whether to survey entire portfolios in one go, or to perhaps undertake surveys on a rolling programme on a five-year basis, will depend on many factors, not least the funds available to commission the survey in the first place. Arguably, mass condition surveys of entire portfolios required in a short period may suffer in terms of quality, but if carried out correctly will give an earlier fuller picture than a rolling programme.

Technology v traditional

Frequently hand-held computer technology is used for condition surveys. Advantages are that this should reduce the time taken to complete the survey as there is no subsequent manual inputting of written notes collected on site, with possible errors being introduced. However, it is important that the software system is correctly designed to suit the portfolio being surveyed. It can be argued that the selection of options available to surveyors on handheld devices may be limited and forces surveyors to select an answer that may not entirely describe the actual situation on site. With the data being downloaded directly into the database, there may also be little chance to review the choices made on the survey and to review the overall approach to the survey on completion when a fuller appreciation of the building and it's faults is available. It is important, therefore, for surveyors to be confident when using technology that it does not undermine their professional competence.

Further reading

Royal Institution of Chartered Surveyors Guidance Note - Stock Condition Surveys

British Standard 8210 (1986) British Standard Guide to Building Maintenance management.

'Best value' in local authorities

What is 'best value?'

The government has defined best value as "a duty to deliver services to clear standards – covering both cost and quality – by the most economic, efficient and effective means available". This clearly has a direct bearing on construction related services provided by local authorities

The government will require local authorities to publish an annual best value performance plan (BVPP) covering the entire range of the authorities' services. This is intended to be a public document and will include an assessment of the authorities' past and current performance against nationally and locally defined standards.

The performance plan will be the main instrument by which local authorities will be held accountable to the local community for delivering best value.

Maintenance management

How are best value studies carried out?

The Audit Commission appoint an external auditor to undertake an audit of the authorities' BVPP. In the case of construction related services this is likely to be a professional consultant in the relevant field. The audit will cover:

- ❖ compliance with legislation and guidance;
- ❖ performance information; and
- ❖ continuous improvement strategies (The 4 C's - see below).

'The 4 C's'

Authorities are required to carry out a Best Value Review as part of the BVPP this includes putting in place strategy for continuous improvement implementing the '4 C's':

- ❖ **Challenge** – why and how the service is being provided.
- ❖ **Compare** – with others' performance (including organisations in the private sector) across a range of relevant indicators.
- ❖ **Consult** – with local taxpayers and service users with the view to setting new performance targets.
- ❖ **Compete** – as a means of securing efficient and effective services.

(More detail regarding the methods of inspection can be seen on the Audit Commission website – www.audit-commission.gov.uk or www.local-regions.detr.gov.uk)

Continuous improvement for construction and property services

A Service Improvement Plan (SIP) is one which identifies areas of potential improvement from:

- ❖ those services already subjected to a Best Value Review;
- ❖ advice from the Audit Commission inspection service; or
- ❖ advice from an external consultant by way of a best value 'healthcheck' (commonly termed a 'critical friend').

Depending on the particular service, the following core issues will be included in the SIP:

- ❖ Description of action – to be taken to improve service.
- ❖ Target – either qualitative or quantitative.
- ❖ Target and timescales – timescale for achieving the improvement(s).
- ❖ Likely effect – impact of the improvement identified.

Other issues which are likely to be included in construction related services could be:

- ❖ current performance;
- ❖ national benchmarks;
- ❖ resource requirements; and
- ❖ monitoring and review strategies.

This overview of best value is intended to give some insight into government efforts in improving services to the taxpayer. When considering the detail of Service Improvement Plans for construction and property related services it is recommended that consultancy advice be sought.

Sources of information in maintenance management

BMI

Building Maintenance Price Book, published annually

BSI

Standards 3811 (Glossary); 3443 (Guide); 6150 (code for decorating) 6270 (cleaning); etc.

BSRIA

Application Guide 1/87.1

Operating and Maintenance Manuals for Building Services Installations 1990

Application Guide 4/89.2 Maintenance Contracts for Building Engineering Services second edition 1992

Application Guide 24/97 Operation and Maintenance Audits 1997

Application Guide 1/98 Maintenance Programme Set-up 1998

Application Guide 20/99 Cost Benchmarks for the Installation of Building Services Parts 1-3 1999

Application Guide 4/2000 Condition Survey of Building Services 2000

NJCC

Procedure Note 16 Record Drawings and Operating and Maintenance Instructions and the health and safety file. Second Edition June 1996 RIBA Publications

Note: NJCC no longer in existence, but guidance notes still in general use

RICS

Building Maintenance: Strategy, Planning and Procurement – A Guidance Note, March 2000, RICS

Computerised Maintenance management Systems – A Survey of Performance Requirements, May 1999, RICS Research paper

Miscellaneous

Building Maintenance management, Chanter & Swallow, second Edition August 2000, Blackwell Science

Building Maintenance and Preservation, Mills 1996, Butterworth-Heinemann Ltd.

Lee's Building Maintenance management, P Wordsworth, fourth Edition November 2000, Blackwell Science

Maintenance management

British standards in maintenance management

BS 3811	Glossary of Terms in Terotechnology	1993
BS 3843	Guide to Terotechnology	1992
	Pt2 Introduction to techniques and applications. Pt 3 Guide to the available techniques	1992
BS 6l50	Code of Practice for Painting Buildings	1991
BS 6270	Code of Practice for Cleaning and Surface Repair of Buildings Pt 3 metals (cleaning)	1991
BS 7543	Guide to Durability of Buildings and Building Elements Products and Components. Still current but partially replaced by:	1992
BS ISO 15686	Buildings and constructed assets. Service life planning. General principles	2000
	Pt2 Service Life Prediction Procedures	2001
	Pt 3 Performance Audits and Reviews	2002
BS 8210	Guide to Building Maintenance management	1986
BS 8221	Code of Practice for Cleaning and Surface Repair of Buildings	
	Pt 1 Cleaning of natural stones, brick, terracotta and concrete	2000
	Pt 2 Surface repair of natural stones, brick and terracotta	2000

Cost management and taxation

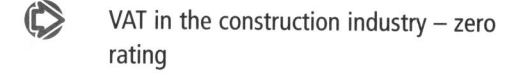

- VAT in the construction industry – zero rating
- Capital allowances for taxation purposes
- Tender prices and building cost indices
- Cost management
- Life cycle costings
- Reinstatement valuations for insurance purposes

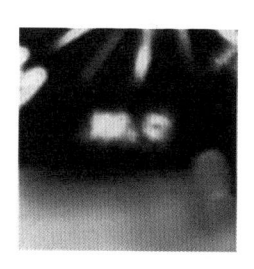

Cost management and taxation

VAT in the construction industry – zero rating

VAT and construction works

VAT in relation to buildings and construction is a complex area but below are some of the fundamental issues to consider when building new or carrying out works on an existing property:-

Building new property

All goods and services supplied by the industry for use in the construction of a building are standard rated (at 17.5% generally and 5% for fuel and power for a qualifying use) except in the following circumstances, when they would be zero rated:-

Where the new building is a dwelling fulfilling the following criteria:-
- ❖ Self contained.
- ❖ Able to be sold as a single dwelling.
- ❖ Has been granted planning consent.
- ❖ Entitled to be used as a dwelling throughout the year.

NB: Certain elements within a dwelling will always be standard rated (see later note). This zero rating only applies to goods and services supplied in the course of construction of a new building and to the first grant of a major interest, by the developer.

Relevant residential building fulfilling the following criteria:-
- ❖ Some facilities shared by residents, for example, children's homes, old people's homes, hospices, living quarters for school pupils or armed forces.
- ❖ An institution which is the sole or main residence for 90% of its residents.

NB: Flats and sheltered housing schemes made up of individual flats will be zero rated as dwellings rather than relevant residential buildings.

NB: Prisons, hospitals and hotels are specifically excluded from zero rating. This applies to goods and services and the first grant of a major interest.

Relevant charitable building – a building used solely by a charity for the non business use of the charity or as a village hall.

NB: The supply must be made to the person who intends to use the building for such purposes before the supply is made then the person receiving it must give to the person making the supply a certificate that the intended use is for relevant residential or charitable purposes.

Works to an existing property

All goods and services supplied will be standard rated except in the following circumstances:-

Relevant Housing Association (HA) converting from non residential to residential use, ie the HA must:
- ❖ be a Registered Social Landlord within the meaning of Part 1 of the Housing Act 1996; and
- ❖ be a registered HA within the meaning of the Housing

Cost management and taxation

Associations Act 1985 (or Part II of the Housing (Northern Ireland) Order 1992).

NB: This applies to goods and services only.

Approved alterations to existing protected buildings – basic principles as follows:-

- ❖ The work must be to a protected building as defined in VAT law and to the fabric of that building (for example, a listed building or scheduled monument).
- ❖ The work must require and be granted listed building consent.
- ❖ The works must not be of a repair and maintenance nature.
- ❖ The works must be apportioned if there is an element of both repair and maintenance and approved alteration.

NB: Unlisted buildings in conservation areas do not qualify as protected buildings for the purposes of VAT relief. This applies to goods and services only.

Substantial reconstruction to existing protected buildings – basic principles as follows:-

- ❖ Definition of protected buildings as above.
- ❖ The work must require and be granted listed building consent.
- ❖ At least 60% of the cost is attributable to 'approved alterations' or
- ❖ Only the wall(s) remain along with the other features of architectural or historic interest.

NB: This applies to the first grant of a major interest to the person carrying out the substantial reconstruction.

Conversion of buildings from non-residential use to dwellings or relevant residential use:-

- ❖ From 1 March 1995 if you are the person converting a building the first grant of a major interest can be zero rated provided the building was neither designed nor adapted as a dwelling (or relevant residential purpose), or if it was designed as a dwelling and/or subsequently adapted, it has not been used as a dwelling since 1 April 1973. See 'New legislation' section below for further changes.

Residential conversions

From 12 May 2001 a lower rate of 5% applies with effect to the following supplies of building services and related goods:-

- ❖ A 'changed number of dwellings conversion', ie a conversion of a building (or part of a building) so that after conversion it has a different number of single-household dwellings (SHD) from the number before the conversion.
- ❖ A 'house in multiple occupation conversion', ie a conversion of a building (or part of a building) containing one or more single-household dwellings so that it contains only one or more multiple-occupancy dwellings.
- ❖ A 'special residential conversion', ie a conversion of premises containing dwellings (single-household or multiple occupancy) for use solely for a relevant residential purpose or converting a care home into a single-household dwelling.

Renovation and alteration of dwellings

The lower rate of 5% also applies to supplies of:

- ❖ building services; and
- ❖ related goods.

In the course of the alteration (including extension) or renovation of a

Cost management and taxation

single-household dwelling that has been empty for three years. 'Empty' means unlived in, so use for another purpose, such as storage, is acceptable. The dwelling can remain a single household.

Specific conditions apply to both of these lower rate situations.

The reduced rate of VAT

The reduced rate of 5% VAT applies from the 1 June 2002. The reduced rate is extended to the costs of:

- ❖ converting a non-residential property into a care home or multiple occupancy dwelling eg bedsits;
- ❖ converting a building used for 'relevant residential' purpose into a multiple occupancy dwelling;
- ❖ renovating or altering a care home or other qualifying building that has not been lived in for three years or more; and
- ❖ constructing, renovating or converting a building into a garage as part of the renovation of property that qualifies for the reduced rate.

Charity annexes

Prior to 1 June 2002 all of an annexe had to be used or intended for use for a relevant charitable purpose in order to zero rate its construction. From 1 June 2002, Customs have issued a concession whereby minor non-qualifying use for example, use where the annex or part of an annexe is not used solely for a relevant charitable purpose, can be ignored with there being no requirement to apportion these between relevant charitable purpose and alternative usage.

Supplies to handicapped people

Certain goods and services supplied to a handicapped person, or to a charity for making these available to handicapped people for their domestic or personal use, may be zero-rated.

DIY projects

For claims made on or after 29 April 1996, Customs and Excise will refund any VAT chargeable on the supply, acquisition or importation of any goods used in connection with construction or conversion work where the following criteria are met. The work is not part of a business project; the work comprises the construction of a dwelling; relevant residential or charitable building; or is the conversion of a non-residential building into a dwelling.

All works, which fulfil the criteria set out above, can be zero-rated. However there are some items, within these categories, which will always be standard-rated:

- ❖ site investigations;
- ❖ temporary site fencing;
- ❖ concrete testing;
- ❖ site security;
- ❖ catering;
- ❖ cleaning to site offices;
- ❖ temporary lighting;
- ❖ transport and haulage to and from site;
- ❖ plant hire (without operator);

Cost management and taxation

- professional services (architects, engineers, surveyors, solicitors etc);
- landscaping;
- furniture (other than fitted kitchens) – material only;
- some electrical or gas appliances – material only; and
- carpet and carpeting materials (including underlay and carpet tiles) – material only.

See the following notices issued by HM Customs and Excise for further expanded information and definitions:

- Notice 708 VAT Buildings and construction
- Notice 742 Land and property
- Notice 719 VAT refunds for 'do it yourself' builders and converters
- Notice 701/1 Charities
- Notice 701/7 VAT reliefs for people with disabilities
- Notice 701/19 Fuel and power
- Notice 708/5 Registered social landlords (housing associations etc)

The law relating to this subject includes:-

- Group 5 of Schedule 8 to the VAT Act 1994 as amended by Statutory Instrument 1995 No 280 and Statutory Instrument 1997 No 50 (reproduced Appendix F) – buildings under construction
- Group 6 of Schedule 8 to the VAT Act 1994 as amended by Statutory Instrument 1995 No 283 (reproduced Appendix G) – approved alterations to protected buildings
- Group 12 of Schedule 8 of the VAT Act 1994 – supplies to the handicapped
- Section 35 of the VAT Act 1994 – DIY scheme
- The VAT (Input Tax) Order 1992 as amended by Statutory Instrument 1995 No 281 (reproduced Appendix H)
- Schedule 10 to the VAT Act 1994 (reproduced Appendix J)

Capital allowances for taxation purposes

The following is intended to provide only a brief overview of the current capital allowances regime in respect of property investment expenditure in the UK.

Property expenditure can broadly be categorised as either for investment purposes or as trading stock (generally for property developers) and as either capital or revenue in nature.

Capital allowances are available on capitalised expenditure incurred on property held for investment purposes.

General principles of capital allowances

Property investors can claim the available tax relief on their property expenditure where the property is held as an investment and the investor is within the charge to tax. Capital allowances are available to investors liable to corporation tax and income tax.

Cost management and taxation

In overview, they can be utilised by the property investor to reduce their end of year tax liability. In effect, they can dramatically increase the yield of a property investment through increased tax efficiency.

They are available to both property owners and occupiers. Many investors, particularly occupiers of leased premises, still do not maximise their entitlement to allowances and are missing valuable tax relief.

Capital allowances are available on all forms of property expenditure including property acquisitions, developments, refurbishments, and fit outs.

General scheme of allowances

For property investment in the United Kingdom, capital allowances are available in the following forms:-

- Machinery or plant allowances – available at 25% per annum on a reducing balance basis of the qualifying expenditure. Expenditure incurred on air conditioning systems, lifts, fire alarm systems and other mechanical and electrical assets will qualify for MPA's.
- Industrial buildings allowance – available at 4% per annum on a straight line basis for qualifying expenditure. This will only apply where the property is in a qualifying industrial use in accordance with current legislation.
- Hotel allowances – available at 4% per annum on a straight line basis for qualifying expenditure. This will only apply where the hotel meets the requirements of current legislation.
- Agricultural buildings allowance – again available at 4% per annum on a straight line basis for qualifying agricultural property.
- Research and development allowances – available at 100% of the total qualifying expenditure in the year in which it is incurred on the provision of premises for carrying out qualifying research and development activities.
- Enhanced capital allowances – available at 100% in the year of incurring the expenditure. ECA's are available for expenditure incurred on energy efficient machinery or plant as detailed by the ECA Technology List issued by the secretary of state.
- Long life assets – available at 6% per annum and applicable to machinery or plant assets which have an expected useful economic life of over 25 years. Exclusions apply for property expenditure incurred on a dwelling house, hotel, office, retail shop and showroom.
- Small or medium sized enterprises – qualifying machinery or plant expenditure incurred by small or medium sized enterprises will qualify for a 40% First Year Allowance (in place of the usual 25%). Where the qualifying assets are for use solely in N Ireland and the expenditure is incurred before 11 May 2002 the First Year Allowance will be 100%.
- Small enterprises – where qualifying IT expenditure is incurred before 31 March 2003 by a small enterprise this will qualify for a 100% First Year Allowance.

Qualifying expenditure

The level of and type of available allowances will vary according to the type and specification of the property and the nature of the expenditure incurred.

The following table provides a general rule of thumb guide for the likely

Cost management and taxation

level of machinery or plant allowances that may be recovered according to given expenditure scenarios.

Expenditure scenario	Qualifying expenditure %
Retail acquisition	2 – 10%
Industrial acquisition	5 – 12%
Office acquisition	14 – 25%
Property refurbishment	40 – 65%
Property fit out	60 – 100%

Additional forms of allowances and revenue deductions may be available subject to property type and qualifying use and the nature of the expenditure incurred.

Property disposals

With recent legislation changes, capital allowances are now an important consideration on property disposals.

Capital Allowances Act 2001 s.198 provides a mechanism for the parties to the contract to agree and fix the level of allowances available to the purchaser on completion of the sale.

Generally, where the vendor has claimed allowances this agreement, or 'joint election', will be made at the vendor's tax written down value (or residue of allowances) on completion.

The effect of this election will be to transfer the remaining allowances to the purchaser. In doing so, this will also protect the vendor's position and prevent them suffering any future 'claw back' of the allowances claimed by the Inland Revenue.

The parties have two years following completion of the sale within which to enter into this election.

Property vendors' and purchasers' should ensure that the capital allowances aspect of a property transaction is researched during the sale negotiations.

Summary

The above is intended as a general overview of a selection of the current provisions for capital allowances in respect of property expenditure.

Readers should be aware that each situation will need careful consideration and application of current legislation and practice. Specialist advice should be sought by the property investor.

Further sources of information

Capital Allowances Act 2001

Income and Corporation Taxes Act 1988

Finance Act 2001

Cost management and taxation

Tender prices and building cost indices

Tender Prices and Building Cost Indices

= Firm
= Forecast

BCIS General Building Cost Index

BCIS All-in Tender Price Index

Index (Base 1985 = 100)

Date

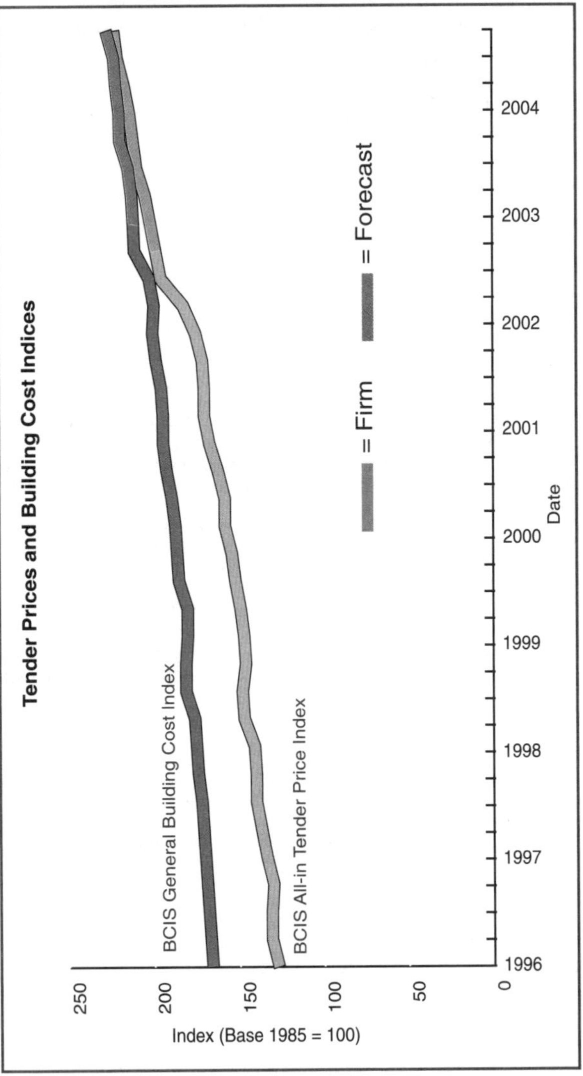

For further information on the breakdown of the indices, please contact the Building Cost Information Service.

Cost management and taxation

The BCIS tender price indices and general building cost indices monitor the movement of tender prices and building costs. They can be used to forecast cost movement for the years to come by using the forecast indices based upon cost trends. The Tender Price Index measures the trend of contractors pricing levels in accepted tender for new work (cost to client), whereas the General Building Cost Index measures changes in costs of labour, materials and plant (cost to contractor).

Cost management

Predicting costs

Be mindful:

- of the level of information;
- of the perceived requirements for budget, time and quality;
- that the brief will inevitably change;
- that not all of your assumptions will be right; and
- that the initial assessment will stick in the client's mind.

Always:

- undertake a detailed assessment;
- set down clearly all assumptions and exclusions; and
- make provision for risk – in the market, in design development and in the client changing his/her mind.

Never:

- rely solely on blanket unit rates; or
- expect the client to remember anything other than your first assessment.

The procurement process

Be mindful:

- that the process of project delivery is dynamic; and
- one of the inevitability of the design (in part at least) developing apace with construction and that if this design development is not managed, cost and time over-runs will result.

Always:

- select an appropriate form of contract and consider not just the value but the nature of the work;
- understand the budget and know where the uncertainties are;
- maintain projections of cost on an 'open book' basis;
- make the projections 'real time' and as soon as issues are suspected, make provision; and
- assess and reassess risk.

Never:

- avoid routine financial appraisals;
- proceed on the basis that things will turn out all right if left to their own devices – they never will; or
- proceed without adequate contingency. To do so is professionally dishonest.

Cost management and taxation

Value for money

Be mindful that:

- value for money is not just the lowest price;
- it must balance time, cost and quality; and
- simply applying competitive tender procedures does not itself demonstrate value for money!

Life cycle costings

> Life cycle costing is defined as the present value of the total cost of an asset over its operating life including initial capital cost, occupation costs, operating costs, and the cost or benefit of the eventual disposal of the asset at the end of its life.

Use

Life cycle costing is essential to effective decision-making in four main ways:-

- It identifies the total cost commitment undertaken in the acquisition of any asset.
- It facilitates an effective choice between alternative methods of achieving a stated objective.
- It is a management tool that details the current operating costs of assets.
- It identifies those areas in which operating costs might be reduced.

It is becoming increasingly important for surveyors in the building industry to offer total cost advice and become proficient in the life cycle cost methods on which such advice is based. Clients should not only be made aware of initial costs such as site costs, capital cost and professional fees, but also of the long term costs of building (for example, occupancy costs, furnishings, operating costs, maintenance costs, energy costs, etc). Such costs can far outweigh initial costs and should therefore have a much stronger influence on building design decisions than is currently the case.

If life cycle cost techniques are to be effective they must be implemented as early as possible in the design process.

Implementing life cycle costings

These must take into account not only the client type but also the investor motivation and each analysis will thus be fashioned to suit the client's particular needs.

Consider the following major elements:-

- The overall time period.
- All costs and revenues attributable to the project, including initial investment, recurring costs and revenues and proceeds from ultimate sale or other disposal.
- Only those costs and revenues directly attributable to the project.
- The effects of time, including allowance for the impact of inflation.
- The fact that pounds spent or received in the future are worth less than pounds spent or received today.

Cost management and taxation

These elements can be broken into seven basic steps necessary in implementing a life cycle cost approach:-

Step 1 – Establish the objective.

Step 2 – Choose a method for achieving the objective (consider all realistic possibilities).

Step 3 – Formulate assumptions (for example, forecast escalation of energy, labour and material costs).

Step 4 – Identify the costs and the life cycle.

Step 5 – Compare costs and rank the alternatives.

Step 6 – Carry out a sensitivity analysis (ie, when the results of step 5 are not demonstrably in favour of one choice).

Step 7 – Investigate capital cost constraints.

The life cycle

Generally, the economic or functional life of the building is chosen as the life cycle but this of course will vary with the type of client. It is probably better to err on the conservative side when forecasting this.

Components of a life cycle cost

Life Cycle Cost Planning (LCCP)
Objectives:-

- ❖ To identify the total costs of the acquisition of a building or building element.
- ❖ To facilitate the effective choice between various methods of achieving a given objective.

Full Year Effect Costs (FYEC)
Objective:-

- ❖ To identify the short term running costs of a proposed building (one to three years).

Life Cycle Cost Analysis (LCCA)
Objectives:-

- ❖ To establish where performance differs from the LCCP projection.
- ❖ To make recommendations on more efficient utilisation of the building.
- ❖ To provide information on asset lives and reliability factors for accounting purposes.
- ❖ To assist in the establishment of a maintenance policy for the building.
- ❖ To give taxation advice on building related items.

While each component can be viewed as a separate activity in its own right, there is a logical sequence that links them together. This can be seen in figure 1 where the assumption is that data for an LCCP of a proposed building is based upon three similar buildings for which LCCA's are available.

Cost management and taxation

The sequence linking LCCA, LCCP and LCCM

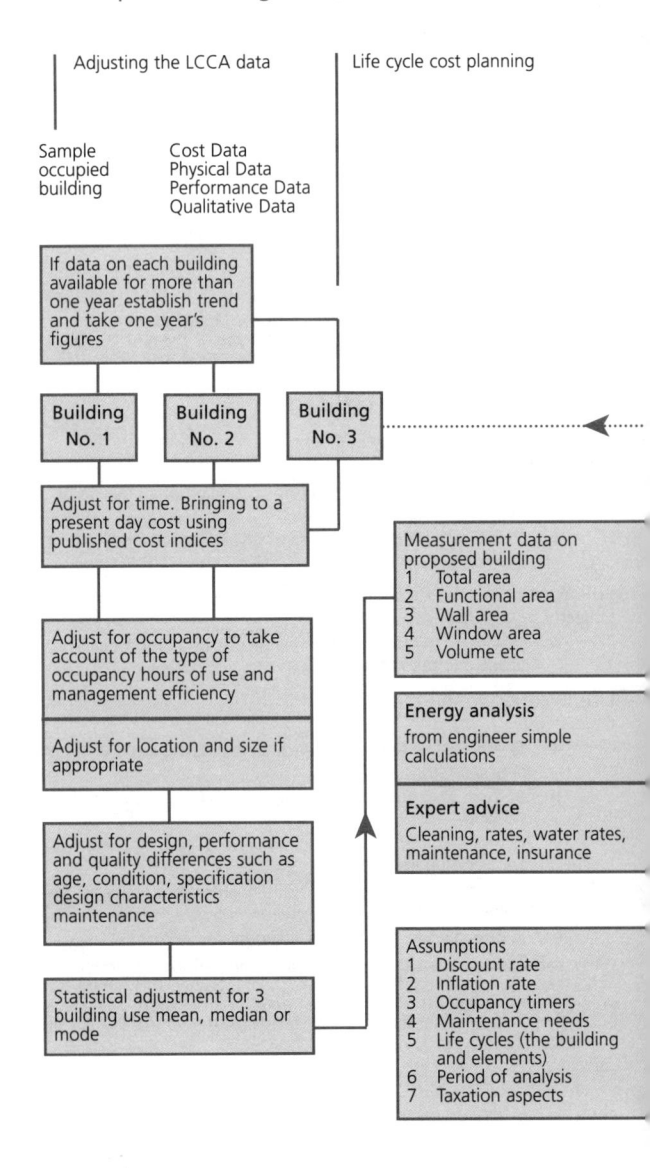

Adjusting the LCCA data | Life cycle cost planning

Sample occupied building

Cost Data
Physical Data
Performance Data
Qualitative Data

If data on each building available for more than one year establish trend and take one year's figures

Building No. 1 | **Building No. 2** | **Building No. 3**

Adjust for time. Bringing to a present day cost using published cost indices

Adjust for occupancy to take account of the type of occupancy hours of use and management efficiency

Adjust for location and size if appropriate

Adjust for design, performance and quality differences such as age, condition, specification design characteristics maintenance

Statistical adjustment for 3 building use mean, median or mode

Measurement data on proposed building
1. Total area
2. Functional area
3. Wall area
4. Window area
5. Volume etc

Energy analysis
from engineer simple calculations

Expert advice
Cleaning, rates, water rates, maintenance, insurance

Assumptions
1. Discount rate
2. Inflation rate
3. Occupancy timers
4. Maintenance needs
5. Life cycles (the building and elements)
6. Period of analysis
7. Taxation aspects

Cost management and taxation

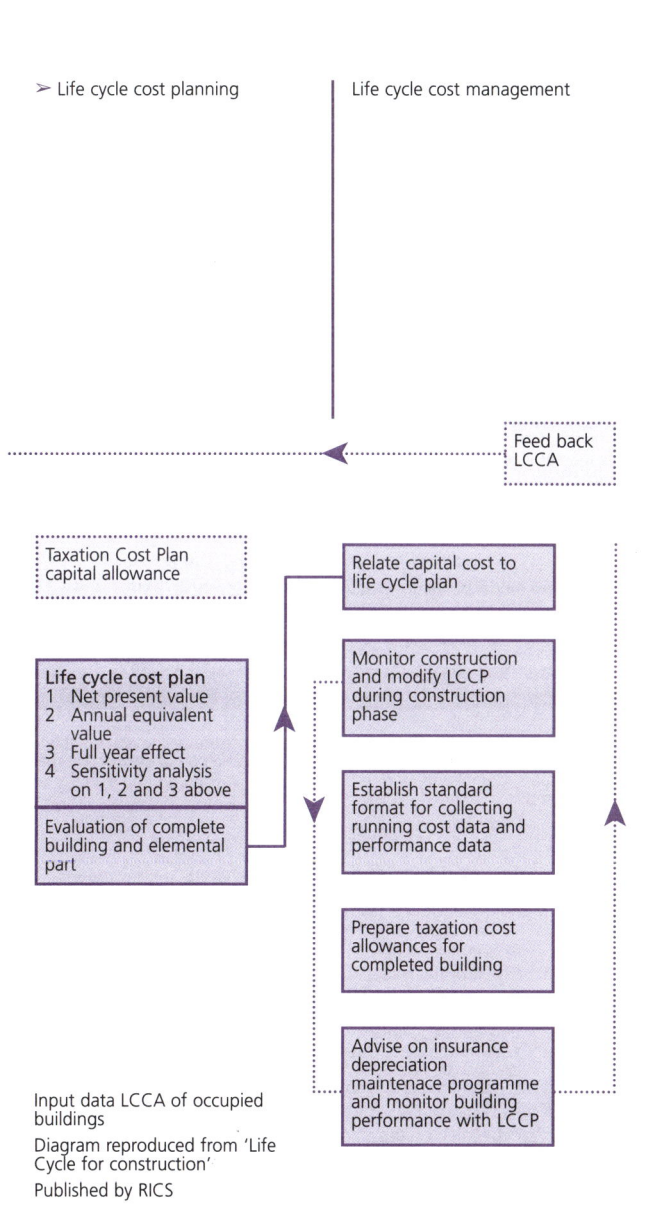

```
➢ Life cycle cost planning        │  Life cycle cost management
                                  │
                                  │
                                  │
                                  │
                                  │
                                  ▼
  ·····································◄─────  ┌──────────────┐
                                               : Feed back    :
                                               : LCCA         :
                                               └──────────────┘
```

```
┌────────────────────┐      ┌──────────────────────────┐
: Taxation Cost Plan :      │ Relate capital cost to   │
: capital allowance  :      │ life cycle plan          │
└────────────────────┘      └──────────────────────────┘

┌────────────────────┐      ┌──────────────────────────┐
│ Life cycle cost plan│     │ Monitor construction     │
│ 1 Net present value │     │ and modify LCCP          │
│ 2 Annual equivalent │     │ during construction      │
│   value             │     │ phase                    │
│ 3 Full year effect  │     └──────────────────────────┘
│ 4 Sensitivity analysis│
│   on 1, 2 and 3 above│    ┌──────────────────────────┐
├────────────────────┤      │ Establish standard       │
│ Evaluation of complete│   │ format for collecting    │
│ building and elemental│   │ running cost data and    │
│ part                │     │ performance data         │
└────────────────────┘      └──────────────────────────┘

                            ┌──────────────────────────┐
                            │ Prepare taxation cost    │
                            │ allowances for           │
                            │ completed building       │
                            └──────────────────────────┘

                            ┌──────────────────────────┐
                            │ Advise on insurance      │
                            │ depreciation             │
                            │ maintenace programme     │
                            │ and monitor building     │
                            │ performance with LCCP    │
                            └──────────────────────────┘
```

Input data LCCA of occupied buildings

Diagram reproduced from 'Life Cycle for construction'

Published by RICS

Cost management and taxation

Main components of discounting

The time stream of costs and revenues
The present and future revenues associated with a particular option should be identified and must take into account the impact of taxation and investment incentives.

The discount rate
Projects should be assessed on the basis of their present values using a discount rate equal to the real (net of inflation) long term cost of borrowing money in the market place. The discount rate will vary with the source of funding.

Project life
An estimate of the probable life of the project should be made. When ranking projects with identical lives, the one with the lowest present value should be chosen, but when they have different lives the cost of each alternative should be expressed as an annual equivalent and the one with the lowest annual equivalent chosen.

Inflation
Present value calculations must take account of inflation. As a guide to a suitable test discount rate, in the present era of low inflation, HM Treasury is recommending the use of a 6% for use in public sector appraisals.

NB: When assumptions or estimates are felt to be uncertain, these should be subjected to a sensitivity analysis.

Data sources
Data from specialist manufacturers, suppliers and contractors.

Data from model buildings. Historical data (for example, BMI).

Reinstatement valuations for insurance purposes

Generally it is wise to insure a property for the value of complete reinstatement following total destruction (including partial destruction that necessitates demolition and rebuild). The following points should be considered:-

- Demolition cost to be valued.
- Rebuilding costs to be valued (initially at 'Day One' rates). Care must be taken here to include for any special features that are to be included to enable replacement of the property on a like for like basis. Examples include façade treatments such as stonework embellishments or internal features such as specialist decorations.
- Extent of the valuation to be discussed and agreed with the client: whether tenants fixtures and fittings are to be valued; whether a basic landlord fit out is to be included; or a shell only valuation with all other items to be included under separate policies. This may be done on agreement with the client or by reference to leases which state the obligations of the Landlord and Tenant.
- Professional fees to be included at an appropriate level, depending on the complexity and location of the property.
- Geographical location factors to be accounted for either in re-building rates or separately by reference to recognised indices.

Cost management and taxation

- ❖ Local authority planning and building regulation fees must be included as they are an unavoidable expense.

If the valuation is being projected to cover the period of the policy (as opposed to Day One valuation) then account should be made for the:

- ❖ period of the policy;
- ❖ design period;
- ❖ planning period;
- ❖ construction period; and
- ❖ void (letting) period (if required by the owner).

This may result in projecting costs for anything up to and beyond a three year period. In addition, as with any construction scheme, a contingency may be included.

In summary, a Day One valuation may increase by upwards of 30% on the basic rebuild costs to take account of the issues listed above. Property owners should always seek a realistic valuation of their property to avoid being over or under-insured. At best the premiums may be based on an exaggerated value and at worst, following destruction, there will be insufficient funds to reconstruct.

Useful information

- Useful information sources

- Useful website addresses

- Contributors to the Watts Pocket Handbook

- Watts and Partners' publications

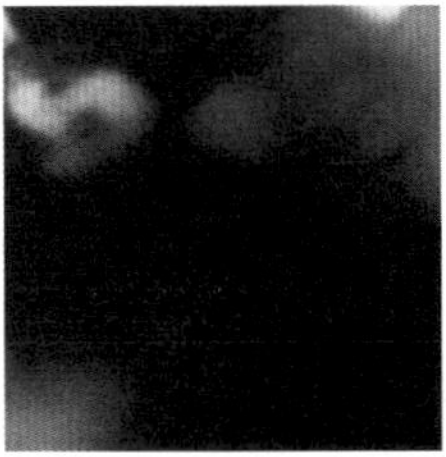

Useful information

Useful information sources

Association for Project Management
Tel: 08454 581944 Web: www.apm.org.uk

Association of Building Engineers
Tel: 01604 404121 Web: www.abe.org.uk

Association of Corporate Approved Inspectors
Tel: 01435 862487 Web: www.acai.org.uk

Building Cost Information Service Ltd
Tel: 020 7222 7000 Web: www.bcis.co.uk

Building Maintenance Information
Tel: 020 7695 1516 Web: www.bcis.co.uk

Building Research Establishment
Tel: 01923 664000 Web: www.bre.co.uk

British Cement Association
Tel: 01344 762676 Web: www.bca.org.uk

British Institute of Facilities Management
Tel: 01799 508608 Web: www.bifm.org.uk

British Property Federation
Tel: 020 7828 0111 Web: www.bpf.org.uk

British Standards Institution
Tel: 020 8996 9000 Web: www.bsi-global.com

Building Services Research and Information Association
Tel: 01344 426511 Web: www.bsria.co.uk

Centre for Accessible Environments
Tel: 020 7357 8182 Web: www.cae.org.uk

Centre for Window and Cladding Technology
Tel: 01125 826541 Web: www.cwct.co.uk

Chartered Institute of Arbitrators
Tel: 020 7421 7444 Web: www.arbitrators.org

Chartered Institute of Building
Tel: 01344 630700 Web: www.ciob.org.uk

Commission for Architecture and the Built Environment
Tel: 020 7960 2400 Web: www.cabe.org.uk

Concrete Repair Association
Tel: 01252 739145 Web: www.concreterepair.org.uk/cra

Concrete Society
Tel: 01344 466007 Web: www.concrete.org.uk

Construction Health and Safety Group
Tel: 01932 561871 Web: www.chsg.co.uk

Construction Industry Council
Tel: 020 7637 8692 Web: www.cic.org.uk

Construction Industry Research and Information Association
Tel: 020 7222 8891 Web: www.ciria.org.uk

Design Council
Tel: 020 7420 5200 Web: www.designcouncil.org.uk

Disability Rights Commission
Tel: 08457 622633 Web: www.drc-gb.org

Energy Watch
Tel: 08459 06 07 08 Web: www.energywatch.org.uk

English Heritage
Tel: 08703 331181 Web: www.english-heritage.org.uk

Environment Agency
Tel: 08459 333111 Web: www.environment-agency.gov.uk

Fire Protection Association
Tel: 020 7902 5300 Web: www.thefpa.co.uk

Glass & Glazing Federation
Tel: 020 7403 7177 Web: www.ggf.org.uk

Useful information

Health and Safety Executive
Tel: 08701 545500 Web: www.hse.gov.uk

HM Land Registry
Tel: 020 7917 8888 Web: www.landreg.gov.uk

House Builders Federation
Tel: 020 7608 5100 Web: www.hbf.co.uk

Institution of Civil Engineering Surveyors
Tel: 0161 972 3100 Web: www.ices.org.uk

Institution of Structural Engineers
Tel: 020 7235 4535 Web: www.istructe.org.uk

Joint Contracts Tribunal
Tel: 020 7637 8650 Web: www.jctltd.co.uk

Lead Development Association
Tel: 020 7499 8422 Web: www.ldaint.org

Mastic Asphalt Council
Tel: 01424 814400 Web: www.masticasphaltcouncil.co.uk

National Trust
Tel: 0870 609 5380 Web: www.nationaltrust.org.uk

Office of the Deputy Prime Minister
Tel: 020 7944 4400 Web: www.odpm.gov.uk

RIBA Bookshop
Tel: 020 7251 7180 Web: www.ribabookshop.com

Royal Forestry Society
Tel: 01442 822028 Web: www.rfs.org.uk

Royal Institute of British Architects
Tel: 020 7580 5533 Web: www.architecture.com

The Royal Institution of Chartered Surveyors (RICS)
Tel: 0870 333 1600 Web: www.rics.org

The Royal Institution of Chartered Surveyors in Scotland
Tel: 0131 225 7078 Web: www.rics-scotland.org.uk

RICS Books
Tel: 020 7222 7000: www.ricsbooks.org

Society of Construction Law
Tel: 01235 770606 Web: www.scl.org.uk

Stone Federation Great Britain
Tel: 020 7608 5094 Web: www.stone-federationgb.org.uk

The Society of Expert Witnesses
Tel: 0845 702 3014 Web; www.sew.org.uk

The Stationery Office Bookshop
Tel: 08702 422345 Web: www.clicktso.com

The Survey Association
Tel: 01784 223760 Web: www.tsa-uk.org.uk

Timber Research and Development Association
Tel: 01494 569600 Web: www.trada.co.uk

Timber Trade Federation
Tel: 020 7839 1891 Web: www.ttf.co.uk

Useful website addresses

Barbour Expert-online to the built environment
www.barbourexpert.com
Well-known provider of information services to construction industry professionals and those responsible for health and safety at work.

Useful information

Boundary Problems
www.boundary-problems.co.uk
This web site has been designed to increase understanding of property boundary and related issues. It aims to help minimise the difficulties that have been experienced by so many people who have suffered a neighbour dispute relating to their property boundary.

BRE Certification Ltd
www.brecertification.co.uk
A leading research-based consultancy, certification and testing business – covering the built environment and associated industries.

The British Council for Offices
www.bco-officefocus.com
The British Council for Offices' mission is to research, develop and communicate best practice in all aspects of the office sector.
It delivers this by providing a forum for the discussion and debate of relevant issues.

British Geological Survey
www.bgs.ac.uk
Site covering details of products and services including building stones services for construction and conservation and an address-linked geological inventory.

Building
www.building.co.uk
UK industry magazine covering the whole spectrum of design and construction activity and all project disciplines.

Building Conservation
www.buildingconservation.com
Online information centre for the conservation and restoration of historic buildings, churches and garden landscapes.

Building Control
www.buildingcontrol.org
Provides information on all aspects of building control in England, Wales and Northern Ireland - plus a complete database of all building control offices and officers.

Businessparks.net
www.businessparks.net
Subscription service providing up to date and in depth data on the UK business park market. The database is continuously updated by a specialist research team.

Construction Best Practice Program
www.cbpp.org.uk
The Construction Best Practice Programme provides support to individuals, companies, organisations and supply chains in the construction industry seeking to improve the way they do business: clients, contractors, consultants, specialists, large or small, public or private. The programme offers a range of services, which raise awareness, gain commitment, support action and facilitate sharing.

Construction Europe
www.construction-europe.com
Site for the pan-European magazine for the construction industry including articles from the latest issues, links, a bookshop and events.

Construction Line
www.constructionline.co.uk
Constructionline is the UK's largest register of pre-qualified construction services. It streamlines procedures by supplying the construction industry and its clients with a single national pre-qualification scheme.

Useful information

Construction Plus
www.constructionplus.co.uk
Portal site with links to news, companies and journals across all construction disciplines.

Construction Research and Innovation Strategy Panel (CRISP)
www.crisp-uk.org.uk
CRISP brings together government, clients, industry and the research community to consider research priorities for construction. It has close links with other rethinking initiatives, including the Movement for Innovation, the Housing Forum and the Construction Best Practice Programme.

Construct Sustainability
www.constructsustainably.co.uk
In the increasingly competitive field of construction, sustainability is becoming a differentiator, capable of providing a vital competitive edge. This website is specifically intended to help the smaller constructors compete successfully in this new area alongside the bigger contracting organisations. It provides practical advice and information in answer to frequently asked questions on sustainability.

Corrosion Prevention Association
www.corrosionprevention.org.uk
The CPA's objective is to identify, quantify and communicate the principles and benefits of cathodic protection to those who have an interest in or influence the design, construction, use, maintenance, preservation and investment in reinforced concrete buildings and structures.

Countyweb
www.countyweb.co.uk
County by county guide to total information including a free property-listing service.

Court Service
www.courtservice.gov.uk
The Court Service is an executive agency of the Lord Chancellor's Department, whose purpose is the delivery of justice. The website covers policy, legislation and the Magistrates' courts.

DRA UK Construction Industry Links
www.dragroup.com/links
Database containing UK construction industry related links, categorised in 158 service categories allowing each category to be searched nationally, in 11 regional areas or in 124 postal areas.

Environmental Organisation Web Directory
www.webdirectory.com
Described as the 'Earth's biggest environment search engine'. Lists sites under 30 major headings.

The European Forecasting Group for the Construction Industry
www.euroconstruct.com
Forecasting trends for the construction market in Europe. It includes 19 European economic and technical research institutes. The site includes details of events and publications.

The Housing Forum
www.thehousingforum.org.uk
The purpose of the Housing Forum is to bring together parties involved in the house building supply chain who are committed and ready to become part of a movement for change and innovation in construction and renovation.

Information for Industry
www.ifi.co.uk
An environmental business news briefing, environmental business magazine, environment business directory and a compliance manual are included on this site.

Useful information

InsideEnergy
www.insidecom.co.uk/eibi
An e-journal for energy in building and industry. Includes an update on the latest industry news, features, case histories, a comprehensive product directory, industry events and a forum to express and exchange news on industry topics.

Integrated Facilities Management
www.i-fm.net
An online resource for facilities management professionals.

Landlord Zone
www.landlordzone.co.uk
An online portal for landlords providing free access to information and resources of value to residential and commercial landlords, tenants, letting agents and other property professionals.

National Radiological Protection
www.nrpb.org
Website of the National Radiological Board. The site gives advice and information on Radon protection in the UK.

Networking Independent Commercial Consultants
http://www.nicc.com
This company was launched in 1998 "to provide a rapid response network of experienced construction consultants who effectively use the communication powers of the Internet". It aims to help people dealing with adjudications under the Housing, Grants, Construction and Regeneration Act 1996. It includes their 'Quarterly business review'.

Party`walls.com
www.partywalls.com
Website that provides access to summaries of some important legal cases relevant to party walls.

Pyramus & Thisbe
www.partywalls.org.uk
An organisation for professionals specialising in party wall matters with a database of consultants and FAQ's

The Quango Website
www.cabinet-office.gov.uk/quango
Site that lists quangos by government department or name of organisation and lists the names of people involved.

Radon Centres Ltd
www.radon.co.uk
Organisation that can help with domestic and commercial properties for the sale of Radon detectors, Radon testing surveys and detector placement and installation of Radon reduction systems

The Reinforced Concrete Council
www.rcc-info.org.uk
The Reinforced Concrete Council was formed in 1988 to promote knowledge and understanding of efficient reinforced concrete design and construction. Initially focusing on commercial building frames, its activities are now much broader, encompassing education in RC, tilt-up construction, and general issues with RC design and construction.

Rethinking Construction
www.rethinkingconstruction.org.rc/
Rethinking construction is the banner under which the construction industry, its clients and the government are working together to improve UK construction performance. Rethinking Construction partners aim to showcase innovations in both products and performance through demonstration projects and highlight best practice available within the industry.

Useful information

Service Charges Guide
www.servicechargeguide.co.uk
This guide sets out overall principles for good practice. It is designed to cover commercial property.

Society of Property Researchers
www.sprweb.co.uk
Site for this professional grouping of researchers. Includes details of events, special interest groups and members.

UK Online Government
www.ukonline.gov.uk
A portal site with links to all government online information and services.

UK Taxation Directory
www.uktax.demon.co.uk
An independently compiled catalogue of websites and material of potential interest to tax professionals and others seeking online information on UK tax matters.

Whole Life Cost Forum
www.wlcf.org.uk
The whole life cost forum is a collaborative initiative by all sectors of the construction industry. The forum has been set up to form the UK database of whole life cost and performance data.

Contributors to the Watts Pocket Handbook

External contributors
The scope of the Watts Pocket Handbook would not be possible without the cooperation of external professionals who have dedicated their expertise and valued time to produce specialist information for inclusion in the publication. For this help and support, Watts and Partners would like to express their sincere appreciation to the following:

Building services design Air conditioning systems Data installations Lift terminology Lighting design Plant and equipment **Conservation and the environment** Energy conservation Environment and specification	Keith Crosby Watkins Payne Partnership 51 Staines Road West Sunbury-on-Thames Middlesex TW16 7AH Tel: 01932 781641 Fax: 01932 765590 Email: wpp@wppgroup.co.uk
Conservation and the environment Contaminated land Environment and specification	Sarah Penry Environ UK Ltd 5 Stratford Place London W1C 1AU Tel: 020 7495 0576 Fax: 020 7499 5382 spenry@uk.environcorp.com
Cost management and taxation VAT in the construction industry – zero rating	Adrian Houstoun Kingston Smith Devonshire House 60 Goswell Road London EC1M 7AD Tel: 020 7566 4000 Fax: 020 7566 4010 Email: ajh@kingstonsmith.co.uk

Useful information

Cost management and taxation
Tender prices and building cost indices

Joe Martin
Building Cost Information Service
Royal Institution of Chartered Surveyors
3 Cadogan Gate
London SW1X 0AS
Tel: 020 7695 1500
Fax: 020 7695 1501
Email: bcis@bcis.co.uk

Building and construction regulations
The Provisions of Part II Housing Grants, Construction and Regeneration Act 1996
Legal and lease
Adjudication under the Scheme for Construction Contracts -
how to get started
Dispute resolution
Expert witness

Suzanne Reeves
Wedlake Bell
16 Bedford Street
Covent Garden
London WC2E 8HF
Tel: 020 7395 3000
Fax: 020 7836 9966

Contracts and procurement
Procurement and standard form contracts

Joe Bellhouse
Wedlake Bell
16 Bedford Street
Covent Garden
London WC2E 8HF
Tel: 020 7395 3073
Fax: 020 7395 3004

Town and country planning in England
Appeals and called-in applications
Development applications and fees
Development plans and monitoring
Environmental impact assessments
General Development Order and Use Classes Order
Listed buildings and conservation areas
Planning policy guidance notes and circulars

Malcolm Judd
Malcolm Judd and Partners
70 High Street
Chislehurst
Kent BR7 5AQ
Tel: 020 8289 1800
Fax: 020 8289 1200
Email: malcolm.judd@mjp.uk.net

Site analysis
Measured surveys

Simon Barnes
Plowman Craven and Associates
141 Lower Luton Road
Harpenden
Hertfordshire
AL5 5EQ
Tel: 01582 765566
Fax: 01582 763180
Email: sbarnes@plowmancraven.co.uk

Watts and Partners' contributors

Access agreements	Aidan Cosgrave
Acts of parliament and regulations	Angela Dawson
Airtightness	Trevor Rushton
Asbestos	Paul Winstone
A systematic approach	Michael Ridley
Basic design data	Robert Clements
'Best value' in local authorities	Stuart Russell
BREEAM	Mark Worthington
The Building (Amendment) Regulations 2001	Trevor Rushton

Useful information

Building types	Robert Clements
Capital allowances for taxation purposes	Steve Smith
Chemical and physical testing requirements	Andrew Tee
Chlorides	Andrew Tee
Commercial/industrial surveys	Trevor Rushton
Common defects in commercial properties	Paul Lovelock
Common defects in residential properties	Trevor Rushton
Computer aided design	Richard Wilson
Condition surveys	Steve Brewer
The Construction (Design and Management) Regulations 1994	Paul Winstone
Construction (Health, Safety and Welfare) Regulations 1996	Paul Winstone
Construction management	Michael Ridley
Construction noise and vibration	Aidan Cosgrave
The contract administrator's role	Andrew Gear
Contributors to the Watts Pocket Handbook	Samantha Rumens
Conversion formulae	Angela Dawson
Coordinating project information	Derek Moir
Corrosion of metals	Trevor Rushton
Cost management	Christopher Knott
Curtain walling systems	Andrew Tee, Jane Dalgliesh
Daylight and sunlight	Aidan Cosgrave
Defects in concrete	Trevor Rushton
Deleterious materials	Andrew Tee
Design loadings for buildings	Andrew Tee
The design process	Graeme Lees
Development monitoring	Campbell MacDougall
Different specifications for different contracts	Tim French
Dilapidations	Alex Charlesworth
The Disability Discrimination Act 1995	Jon Hubbard, Trevor Rushton
Discovery of building defects – statutory time limits	Paul Lovelock
Due diligence and the building survey	Trevor Rushton
Employer's agent	Christopher Knott
The Fire Precautions Act 1971	Paul Winstone, Angus Taylor
The Fire Precautions (Workplace) Regulations 1997 and (Amendment) Regulations 1999	Paul Winstone, Angus Taylor
Fungi and timber infestation in the UK	Ian Ford
Glazing - windows and doors satisfying the Building Regulations	Trevor Rushton
Health and safety at work	Paul Winstone
High Alumina Cement concrete	Andrew Tee
Institutional standards	Trevor Rushton
Introduction (chairman's)	Peter Primett
Latent Damage Act 1986	Paul Lovelock
Life cycle costings	Michael Ridley
Mechanisms of water entry	Trevor Rushton
Non-destructive testing	Andrew Tee
Partnering	Christopher Knott
Party wall procedure	Aidan Cosgrave, Paul Lovelock
Primary objectives	Michael Ridley
Problem areas with 1960s buildings	Trevor Rushton
Procurement methods	Andrew Gear
Project management	Michael Ridley
Quality management and professional construction services	Michael Ridley, Aidan Cosgrave
The quantity surveyor's role	Stuart Russell
Radon	Robert White

Useful information

Reinstatement valuations for insurance purposes	Stuart Russell
Residential surveys	Trevor Rushton
Rights to light	Aidan Cosgrave, Paul Lovelock
Rising damp	Trevor Rushton
Rising groundwater	Andrew Tee
Site archaeology	Allen Gilham
Soil survey	Andrew Tee
Soils and foundation design	Andrew Tee
Sources of information in maintenance management	Michael Ridley
Specification writing	Tim French
Specifications	Tim French
Spontaneous glass fracturing	Trevor Rushton
Tender prices and building cost indices	Paul Foley
Useful information sources	Elizabeth Pater, Angela Dawson
Useful website addresses	Elizabeth Pater, Angela Dawson
VAT in the construction industry – zero rating	Stuart Russell
Vendor's surveys	Trevor Rushton
Watts and Partners' publications	Samantha Rumens, Richard Samuel

Watts and Partners' publications

For further information on any of the following publications, please contact Watts and Partners' Marketing Department on +44 (0)20 7090 7820.

Watts Bulletin
A monthly supplement to the Watts Pocket Handbook, the Watts Bulletin keeps its readership abreast of breaking news and changes in legislation within the property and construction industry. It also delivers information on a variety of other topics, including environmental issues, health and safety, new materials and the latest techniques within the industry. In the coming months, Watts and Partners plan to develop a web-based version of the Bulletin for those readers who would prefer this format.

www.WattsandPartners.com
The Watts website holds current and future information about Watts and Partners including the people, the services, the projects, the publications and technical and general news. Planned developments for the website focus around making it more interactive, with a new homepage that shows the latest news, a dynamic recruitment section and an events page.

Watts Review
This annual publication provides a flavour of the firm's activities in the previous 12 months and an overview of its people, personality, values, history and plans for the future. Watts Review 2002, the most recent issue, presents the competitive advantage that a growing office network has given the practice. It identifies the benefits that the diversification throughout UK and Ireland offers clients and demonstrates the firm's scope of service, skillset and experience.

Watts Practice Profile
The Profile provides an account of Watts and Partners' offices, services, performance and procedures, and is presented as a matter of course, when tendering for new business. As a control document for all other marketing publications, it is continually updated.

Index

Index

TITLE	PAGE
1960s buildings	150-158
Above ground archaeology	97, 100
A systematic approach	194-197
Access agreements	89
Acquisition	3-16, 81, 83, 84, 124, 148, 208, 209, 212, 213
Actionable injury	81
Active and inactive systems	177, 178
Acts of parliament and regulations	30-33
Adjudication	45, 46, 48, 75-77
Adjudication under the Scheme for Contruction Contracts - how to get started	76, 77
Air conditioning	42, 68, 116-118, 187, 188, 208
Air conditioning systems	116-118, 208
Air handling units	119
Airtightness	43, 44
Airtightness testing	44
Ancient monument	30, 31, 98-100
Appeals	56, 58, 59
Appeals and called-in applications	56
Arbitration	30, 46, 75
Archaeology	58, 97-100
Architectural and design criteria	91-102
Architectural, engineering and services design	91-132
Asbestos	4, 10, 30, 32, 71, 134, 136-141, 149, 151-153, 171
Basic design data	92
'Best value' in local authorities	199, 200
Bi-metallic corrosion	145
Boilers	119, 187
Boreholes	104, 106, 107, 167
BREEAM	191, 192
British standards	101, 104, 202, 220
Building Act	30, 31, 33
The Building (Amendment) Regulations 2001	30, 42, 43
Building and construction regulations	41-53
Building control	31, 222
Building defects	71, 147-167
Building legislation and control	29-39
Building materials	165, 166, 171, 172, 187, 189
Building regulations	6, 30, 32, 33, 39, 42, 43, 141, 179, 180, 187, 191
Building research	85, 191, 220

Index

TITLE	PAGE
The Building Research Environment Assessment Method	191, 192
Building services	97, 115-122, 148, 187, 197, 201, 205, 220
Building services design	115-122
Building survey	12-14, 148
Building types	93-95, 118, 151, 191
CAD	125, 128-130
Called-in applications	56, 57
Capillarity	178
Capital allowances	23, 30, 207-209
Capital allowances for taxation purposes	207-209
Capital project advice	23
Carbonation	4, 143, 144, 152, 155, 159, 160, 170
CDM	5, 51-53
Cement	4, 135, 141-144, 153, 154, 156, 158, 170, 171, 220
CFCs	6, 119, 188, 189
Chemical and physical testing requirements	170, 171
Chloride Ion Content	170, 171
Chloride induced corrosion	143, 144, 159
Chlorides	143, 144, 165, 170
Circulars	58, 59
Cladding	13, 43, 134, 145, 148, 156, 173-182, 220
Cladding design	177, 178
Commercial buildings	177, 187, 191, 194
Commercial and industrial property	3-6
Commercial/industrial surveys	4-6
Common defects in commercial properties	148
Common defects in residential properties	149, 150
Components of discounting	216
Computer aided design	128-130
Concrete	104, 135, 136, 141-144, 149, 151-160, 170-172, 182, 190, 202, 206, 220, 223, 224
Condition surveys	128, 195, 197-199
Conservation	4, 31-33, 42, 43, 56-61, 98-100, 183-192, 205, 222
Conservation and the environment	183-192
Conservation plan	100

Index

TITLE	PAGE
Construction (Health, Safety and Welfare) Regulations 1996	30, 49-51
Construction Act	74-76
Construction contracts	18, 19, 20, 31-33, 44-48, 75, 76
The Construction (Design and Management) Regulations 1994	51-53
Construction management	18, 19, 26
Construction monitoring	14
Construction noise and vibration	90
Construction projects	18, 101
Contaminated land	5, 10, 30, 33, 137, 184-186
Contamination	9, 12, 105, 107, 108, 119, 134, 144, 151, 165, 170, 184-186
Contract administrator	20, 188, 189
The contract administrator's role	20
Contract management	21-24
Contracts and procurement	17-20
Contributors to the Watts Pocket Handbook	225-228
Control of asbestos at work	30, 32, 137
Control systems	120
Conversion formulae	132
Coordinating project information	101, 102
Corrosion	135, 136, 142-145, 149, 151-157, 159, 165
Corrosion of metals	145, 165
Cost management	23, 203-217
Cost management and taxation	23, 203-217
Curtain walling systems	174-178
Damp	9, 135, 136, 144, 149, 154, 161-166
Data installations	122
Daylight and sunlight	85, 86
Defects in concrete	159, 160
Deleterious materials	6, 12, 14, 134-136, 148, 149
Diminution in value	66, 67, 82
Design and build procurement	18, 22
Design data	92, 130
Design loadings for buildings	108-114
The design process	24, 26, 53, 104, 192, 212
Design specification	96
Desktop studies	98, 99
Development	25-28

Index

TITLE	PAGE
Development and procurement	6, 14-16, 17-28
Development applications and fees	56
Development monitoring	14-16
Development plans	56-58
Development plans and monitoring	56, 57
Different specifications for different contracts	96
Dilapidations	13, 64-70
Disability and the building regulations	39
Disability Discrimination	5, 12, 30, 36-39, 67
The Disability Discrimination Act 1995	30, 36-39, 67-69
Discovery of building defects - statutory time limits	71
Dispute resolution	74-76
Due diligence	11-16, 184
Due diligence and the building survey	12, 13
Employer's agent	22
Energy conservation	4, 32, 43, 186, 187, 198
Environment and specification	188, 189
Environmental impact assessments	61
Equilibrium moisture content	165
Evidence in court	73, 74
Expert witness	64, 71-74
Field evaluation	99
Finance monitoring	15, 16
Fire Precautions	6, 30-36, 93
The Fire Precautions Act 1971	30-32, 34-36, 93
Fire Precautions (Workplace) Regulations	30, 32, 34
The Fire Precautions (Workplace) Regulations 1997 and (Amendment) Regulations 1999	34-36
Fungi and timber infestation in the UK	160-164
Gasket design	179
General Development Order and Use Classes Order	57, 58
Geotechnical	104-107, 185
Glass fracturing	181, 182
Glazing	5, 12, 43, 151, 156, 174-176, 179, 180
Glazing - windows and doors satisfying the building regulations	179, 180
Gravity	178
Groundwater	104-106, 108, 165-167, 184
HAC	133, 135, 141-143, 170
Halons	188
Hazardous to health	30, 32, 48, 134, 162
HCFCs	119, 188, 189

Index

TITLE	PAGE
Health and safety	5, 6, 10, 16, 35, 48-53, 140, 148, 151, 170, 171, 194, 198, 201, 220, 221
Health and safety at work	6, 10, 48, 49, 221
High Alumina Cement concrete	141-143
Housing Act	31, 33, 204
Housing and residential property	7-10
Identification of radon	190, 191
Independent expert	64, 75
Industrial buildings	44, 148, 208
Infestation	10, 149, 160-164
Insect	10, 149, 162, 171
In-situ testing	106
Installations	14, 122, 187, 188, 201
Institutional standards	6
Introduction	1
Kinetic energy	178
Landlord and tenant	31, 65, 66, 69, 70, 195, 216
Landscaping	148, 187, 197, 207
Latent Damage Act 1986	31, 70, 71
Legal and lease	63-77
Legal issues	13, 101
Legionnaires' disease	6
Legislation	29-90, 101, 137, 162, 185-187, 200, 205, 208, 209, 223
Legislation and regulation	29-90
Life cycle costings	212-216
Lift terminology	120, 121
Lighting	5, 12, 35, 42, 50, 82, 85, 86, 92, 121, 122, 187, 206
Lighting design	121, 122
Lighting requirements	92
Listed building	31, 33, 42, 56, 57, 59-61, 100, 205
Listed building consent	56, 60, 100, 205
Listed buildings and conservation areas	31, 59-61
Litigation	64, 75
Local authority	19, 42, 43, 57, 61, 82-85, 90, 93, 94, 217
Maintenance management	193-202
Materials	4, 6, 10, 12, 14, 15, 45, 60, 97, 101, 133-145, 189, 192, 194, 196, 207, 211

Index

TITLE	PAGE
Materials and components	134, 155-184, 194
Materials and defects	133-182
Measured surveys	124, 125, 226
Mechanisms of water entry	178
Mediation	75, 78
Metric conversions	92
Monitoring service	15
Moulds	144, 151, 160, 161, 163
National Disability Council	36, 38, 39
NDT	171, 172
Neighbourly matters	79-90
Nickel sulphide	181, 182
Noise	9, 58, 79, 90, 92, 116-119
Noise and acoustics	92
Noise and vibration	119
Non-destructive testing	171, 172
Northern Ireland Disability Council	39
Notice of adjudication	76, 77
Nuisance	80, 90
Occupied property	68
Offices	31, 33, 34, 92-94, 109, 111, 116, 118, 120, 121, 148, 191, 206, 222
Open book	24, 211
Ozone depletion	188
Partnering	19, 20, 21, 23, 24, 102
Party wall procedure	86-89
Payment	16, 20, 22, 31, 33, 45-47, 68, 72, 82, 89, 186,
Performance related reward	24
Performance specification	96
Planning policy	97
Planning policy guidance notes and circulars	58, 59
Plant and equipment	50, 119
Plaster/mortar	171
Power systems	120
Predicting costs	211
Preservation	60, 61, 98, 99, 201, 223
Pressure differential	175, 178
Primary objectives	194
Problem areas with 1960s buildings	150-158
Procurement and standard form contracts	18-20
Procurement methods	18, 26
Project information	101, 102
Project management	14, 19, 27, 101, 102, 220

Index

TITLE	PAGE
Property acquisition and disposal	3-16
The Provisions of Part II Housing Grants, Construction and Regeneration Act 1996	44-48
Public transport	36, 38
Quality management and professional construction services	27, 28
The quantity surveyor's role	23
Radon	189-191, 224
Reinstatement valuations for insurance purposes	216, 217
Repairs and defects	12
Residential properties	149, 150, 189
Residential surveys	8-10
Right of way	89
Rights to light	13, 80-85, 88
Rising damp	165, 166
Rising groundwater	167
Schedule of Dilapidations	64-70
Scheduled ancient monument	98-100
Site analysis	123-125
Site archaeology	97-100
Soil survey	105-108
Soils and foundation design	104, 105
Sources of information in maintenance management	201, 202
Specification writing	96, 97
Specifications	65, 96, 97, 102, 176, 179
Spontaneous glass fracturing	181, 182
Standard hose test	176
Statutory time limits	71
Stick system	174
Structural and civil engineering design	103-114
Sulphate attack	142, 156
Sulphates in concrete	170
Surface tension	166, 178
Surveys	4-6, 8-10, 14, 105-108, 124, 125, 128, 139, 161, 172, 184, 185, 195, 197-199, 224
Tables and statistics	131, 132
Tender prices	89, 210, 211
Tender prices and building cost indices	210, 211
Testing	169-172
Testing requirements	170-172
Thermal break	175, 179
Thermal insulation	158, 186, 187

Index

TITLE	PAGE
Thermography	172
Timber	104, 149, 155, 160-166, 171, 172, 179, 188, 190, 221
Town and country planning in England	55-61
Two stage tendering	24
Use classes order	56-59
Useful information	219-236
Useful information sources	220, 221
Useful website addresses	221-225
Value for money	22, 23, 194, 196, 212
Variable air volume (VAV)	117
VAT in the construction industry - zero rating	204-207
Vendor's surveys	14
Warehouses	95, 112, 148, 153, 154
Water chillers	119
Water entry	174, 177, 178
Watts and Partners' publications	228
Zero rating	204-207

Watts and Partners

Consultants to the property and construction industry

Belfast
2-12 Montgomery Street
Belfast BT1 4NX
Tel: +44 (0)28 9024 8222
Fax: +44 (0)28 9024 8007

Bristol
33-35 Queen Square
Bristol BS1 4LU
Tel: +44 (0)117 927 5800
Fax: +44 (0)117 927 5810

Dublin
74 Fitzwilliam Lane
Dublin 2
Tel: +353 (0)1 703 8750
Fax: +353 (0)1 703 8751

Edinburgh
County House
63-65 Shandwick Place
Edinburgh EH2 4SD
Tel: +44 (0)131 229 9340
Fax: +44 (0)131 229 9800

Glasgow
176 Bath Street
Glasgow G2 4SE
Tel: +44 (0)141 353 2211
Fax: +44 (0)141 353 2277

Leeds
Atlas House
31 King Street
Leeds LS1 2HL
Tel: +44 (0)113 245 3555
Fax: +44 (0)113 245 1333

London (City)
1 Great Tower Street
London EC3R 5AA
Tel: +44 (0)20 7090 7820
Fax: +44 (0)20 7623 4450

London (West End)
11 Haymarket
London SW1Y 4BP
Tel: +44 (0)20 7533 4800
Fax: +44 (0)20 7839 4740

Manchester
Brook House
77 Fountain Street
Manchester M2 2EE
Tel: +44 (0)161 236 0777
Fax: +44 (0)161 236 0747

www.WattsandPartners.com